Lessons from
a Lean Consultant

Lessons from a Lean Consultant

Avoiding Lean Implementation Failures on the Shop Floor

Chris A. Ortiz

PRENTICE
HALL

Upper Saddle River, NJ • Boston • Indianapolis • San Francisco
New York • Toronto • Montreal • London • Munich • Paris • Madrid
Capetown • Sydney • Tokyo • Singapore • Mexico City

The publisher offers excellent discounts on this book when ordered in quantity for bulk purchases or special sales, which may include electronic versions and/or custom covers and content particular to your business, training goals, marketing focus, and branding interests. For more information, please contact:

U.S. Corporate and Government Sales
(800) 382-3419
corpsales@pearsontechgroup.com

For sales outside the United States please contact:

International Sales
international@pearsoned.com

This Book Is Safari Enabled

The Safari® Enabled icon on the cover of your favorite technology book means the book is available through Safari Bookshelf. When you buy this book, you get free access to the online edition for 45 days.

Safari Bookshelf is an electronic reference library that lets you easily search thousands of technical books, find code samples, download chapters, and access technical information whenever and wherever you need it.

To gain 45-day Safari Enabled access to this book:
- Go to www.informit.com/onlineedition
- Complete the brief registration form
- Enter the coupon code FBM3-4SUM-RFMB-Q1M3-HDBI

If you have difficulty registering on Safari Bookshelf or accessing the online edition, please e-mail customer-service@safaribooksonline.com.

Visit us on the Web: www.informit.com/ph.

Library of Congress Cataloging-in-Publication Data
Ortiz, Chris A.
  Lessons from a lean consultant : avoiding lean implementation failures on the shop floor / Chris A. Ortiz.
      p. cm.
  Includes index.
  ISBN 0-13-158463-4 (pbk. : alk. paper)
  1. Production management. 2. Manufacturing processes. 3. Industrial efficiency. I. Title.
  TS155.0772 2008
  658.5—dc22
                                                                2007051137

ISBN-13: 978-0-13-158463-1
ISBN-10:   0-13-158463-4

Text printed in the United States on recycled paper at Donnelley in Crawfordsville, Indiana.
Second printing, March 2008

*This book is dedicated to my wife, Pavlina,
and my two boys, Sebastian and Samuel,
who support me in my business
and writing endeavors.*

Contents

Preface

By itself, implementing lean manufacturing on the factory floor will not generate the desired results unless certain systems are in place, production support staff operate efficiently, and the operators are fully engaged and committed to the change. As a lean consultant, I have traveled around the country and have seen a variety of companies and manufacturing operations. Each one was unique in its approach to lean manufacturing.

I am often asked why some companies are successful and others fail in their lean implementations. Or more simply put, what did they do to get the results they obtained?

Lessons from a Lean Consultant outlines the fundamental mistakes companies make in trying to implement lean manufacturing. I provide solutions for upper managers, engineers, and supervisors who are struggling to keep their lean implementations afloat or are looking for advice on how to mold their production operators into the powerful change agents they need to be. *Lessons from a Lean Consultant* helps prepare managers for working in a lean shop floor environment. Lean succeeds or fails depending on the commitment from management. Dealing with the ups and downs of lean implementations can be tough. This book shows what is expected of managers and their subordinates.

From my perspective, the lean manufacturing training and consulting business has changed over the years. It used to be common practice for companies to spend tens of thousands of dollars on individuals who taught generalized, theory-based curriculums. I am not implying that all consultants operated in this manner, but many did. This type of training and guidance has become outdated, and companies are looking for hands-on approaches to lean training.

This change in the market has affected my books on lean manufacturing. I saw this market transformation about two years ago and wrote *Kaizen Assembly: Designing, Constructing, and Managing a Lean*

Assembly Line in an attempt to cater to the nuts-and-bolts needs of the market. *Lessons from a Lean Consultant* takes the same approach to lean management and culture change, providing detailed, real-life solutions to those who are struggling with workers who do not want to adapt to lean methods or adhere to the systems and procedures in place. In short, this book takes a nuts-and-bolts approach to lean management and culture transformation.

Chapter 1 briefly tells the story of how an attempt to implement lean manufacturing failed in a company for which I was the senior lean manufacturing engineer. Chapter 2 lays the foundation for the book by outlining the importance of meeting customer needs and developing a competitive balance between cost, quality, and delivery. Chapter 3 provides a step-by-step guideline to creating a company kaizen program, an organization's foundation for continuous improvement. Chapter 4 helps the manufacturing professional identify the early stumbling blocks that commonly occur during the planning and implementation of lean manufacturing in a factory environment. This chapter defines the cornerstones of the early stages of lean manufacturing: 5S, visual management, data collection, waste removal, and process design.

Chapters 5 and 6 illustrate the importance of developing a relationship with your line operators and teaching them how to work and interact in a lean process. Success requires solid up-front planning and training of existing and new employees. Chapter 7 feeds off of Chapters 5 and 6, explaining in detail the incentive and pay-for-skill systems that help you encourage cross-training and develop a flexible workforce.

Chapter 8 takes a unique approach to lean leadership. This chapter emphasizes how the best leaders treat people, with or without a lean journey. In short, if company leaders manage through negative reinforcement, the organization will struggle to adapt to a lean culture.

Appendixes A and B illustrate forms, documents, and templates that are useful during your lean journey. The Glossary lists and defines the key lean terms you should know.

It's my hope that professionals working in the manufacturing sector holding titles such as plant manager, engineering manager, lean manager, and even lean engineer will find value in this book. My goal is to provide a tool for creating a solid lean program in which the people in direct control of its success are driving the improvements, thereby ensuring that their implementations endure and prosper.

Acknowledgments

First and foremost, I have to give thanks to my family—my wife, Pavlina, and my two sons, Sebastian and Samuel. Without their support and encouragement, it would have been nearly impossible for me to devote the time to this book that I did. All of my professional efforts are to ensure our success together.

Second, I want to thank the staff at Prentice Hall who first took interest in this book and their professional approach to making it become a reality. I want to acknowledge my editor, Bernard Goodwin, for his passion for lean manufacturing and for taking on my project. I also have to thank Betsy Hardinger for her support in helping shape the manuscript.

The final name to mention is Debra Riffin. Debra was critical to the completion of this manuscript. I want to thank her for editing and proofreading the manuscript and helping me better convey my thoughts. Debra, you were a valuable asset.

It is important also to recognize the countless number of manufacturing professionals I have had the luxury of working with over the years. My clients are the true experts. I learn from them, and the relationships I have developed with them are critical to my company's success.

About the Author

Chris Ortiz is a senior lean consultant and the owner of Kaizen Assembly. He has spent the majority of his professional career working for *Fortune* 500 companies, teaching and guiding them to become more efficient businesses. Chris has also led more than 150 kaizen events around the United States.

Ortiz is also an instructor at five Washington state community and technical colleges. He has developed a reputation for delivering fast-paced, highly detailed, and interactive classroom-style courses.

He is the author of the book *Kaizen Assembly: Designing, Constructing, and Managing a Lean Assembly Line* (Boca Raton, FL: Taylor and Francis Group, 2006). His lean implementation techniques have been featured in a variety of trade magazines, newspapers, corporate newsletters, *Industrial Engineer* magazine, *Industrial Management* magazine, and other lean manufacturing newsletters and periodicals.

To contact Chris Ortiz, e-mail chrisortiz@kaizenassembly.com or go to his Web site, www.kaizenassembly.com.

Introduction

At first glance, lean manufacturing appears to be a magic solution for the numerous problems experienced in many factories. A powerful and effective improvement philosophy, lean manufacturing can prevent company failure—or catapult a business into the status of a world-class organization. It's well known and widely accepted that many improvements—such as enhanced productivity, improved quality, shortened lead times, and reduced costs—are the direct result of employing lean manufacturing methodologies.

Without a doubt, embarking on a lean journey is a wise choice as well as a bold one. It is important to recognize, however, that not all stories end in success. Although few people in the manufacturing environment realize it, there are many failures and struggles in implementing lean methods on the factory floor.

Admittedly, it's tough to acknowledge failure, but it's the failure of lean implementations that have provided valuable insight regarding what not to do when you're doing lean. The key to success is total commitment to the process, and not only from management. Although management is the main driver for all lean initiatives, success depends on dedicated focus by engineers and supervisors and full engagement by line operators and other product building personnel. However, it is hard to remain dedicated to the effort if errors were made in the implementation or if the appropriate systems are not in place to provide adequate support, monitoring, and the ability to make ongoing improvements.

Lean manufacturing is successful only when the individuals running the business are committed and dedicated. Implementations fall apart because of lack of management support, poor procedures and standards, and lack of accountability and vision. In truth, we must admit that lean manufacturing has failed in many companies, even though the importance of management's commitment to lean is not a new concept. It has been repeated over and over, almost like a broken record, especially when we compare U.S. manufacturers to their competitors in industrialized nations such as Japan, which was the pioneer of what is now known as lean and kaizen.

This fact by no means diminishes the years of effort in developing methods such as the seven deadly wastes, 5S, kaizen, waste reduction, single piece flow, and visual management. However, it is time to take the concept of lean management down to a level that focuses on the people who can make, or break, an implementation. Workers need a solid understanding of lean manufacturing, and training of all employees involved in manufacturing operations is critical. Typically, companies train only the individuals who are responsible for the initial design and implementation of the lean manufacturing systems and neglect to formally train the line operators, giving them only quick, on-the-fly training that is more likely to confuse and frustrate them than adequately prepare them for the new process.

Plenty of books are available that address the general issues of the struggles of implementing a lean strategy. They provide good direction for creating the overall lean strategy, but what about those who work on the shop floor, day in and day out, during the lean implementation and afterward? How do you deal with operators who do not buy in to new ideas? Why does your production supervisor avoid holding employees accountable for making mistakes and ignoring standard operating procedures? And why doesn't your production manager hold the supervisor accountable? Why did the 5S program fall apart? Most importantly, why has the company been unable to achieve the shop floor performance that was originally estimated? These problems stem from the people factor of a lean implementation. That's what this book addresses.

As a manufacturing professional, you probably know that implementing 5S, single piece flow, visual management, and other lean philosophies will help significantly reduce process inefficiencies. You must conduct solid up-front planning—in the form of training and data collection—and monitor the process closely. Sometimes, only a slight tweak here or there is all you need. Rather than detail the infinite attributes of the lean enterprise and overall change philosophies, this book turns lean management into lean mentoring. It guides you through the critical aspects of early design and data collection and provides clear, simple rules for sustaining the new processes after they are in place.

The most common struggle in a lean journey does not typically occur during the learning phase. I have been a lean manufacturing trainer and consultant for some time now, and we are a dime a dozen. You can find efficiency consultants anywhere. Most manufacturers have come to realize the benefits of a lean approach and have employed the finest talent to implement it. Most of your employees—engineers, technicians, production managers, engineering managers, and plant managers—have the ability to grasp the fundamental concepts of lean manufacturing and can envision the impact it can have on organizational performance and financial strength. Often, what's missing from many lean manufacturing programs is the final piece of the puzzle: a company kaizen program.

If you have read my first book, *Kaizen Assembly: Designing, Constructing, and Managing a Lean Assembly Line,* you know that I am a strong advocate of kaizen (continuous improvement) and the use of kaizen events to implement lean manufacturing in an efficient and organized manner. That is why this book explains in detail the concept of kaizen and kaizen events. Kaizen events themselves are an effective method of accelerating lean implementations. However, it is important not to get stuck in "event lean," in which improvements happen only in scheduled phases. This is why it's crucial to understand the differences between kaizen and kaizen events, something you'll learn in this book.

I've also written this book to help the manufacturing professional identify the early stumbling blocks that commonly occur during the planning and implementation of lean manufacturing in a factory environment. Companies often are eager to adopt techniques such as 5S, visual management, data collection, waste removal, and process design, but they learn and implement them in the wrong manner. For example, many companies achieve the first four S's of 5S (sort, straighten, scrub, and standardize), but the program falls apart because no one is made accountable for the fifth element: sustain.

Here's another example. Gathering current state data is essential to improving performance, but the information is either not available or is collected improperly. To achieve a smoothly run operation, you must accurately identify waste and eliminate it from the process—a difficult task that also can be done incorrectly. In this book I outline the fundamental aspects of 5S, visual management, data collection, waste removal, quality at the source, and workstation design in words that are clear and easily understood.

Other keys to success include installing support systems, ensuring that production support staff operate in an efficient manner, and making certain that operators are fully engaged and committed to the change. To this end, I explain the concept of standard work, an agreed-upon set of work procedures that outlines the best, most reliable, and safest way of doing work for production operators.

Training new and existing employees takes on a whole new meaning in a lean environment. Whether they are hourly operators or salaried support staff, your employees need to understand the fundamentals of lean manufacturing. Operators should no longer be trained simply on how to build product. They must interact in a process that is highly organized and monitored by procedure and structure. I describe how to develop a comprehensive lean training program for new and existing employees as a critical part of any lean organization.

Although temporary workforce layoffs are common after lean manufacturing implementation, truly lean, world-class factories can actually create new positions and jobs. Lean organizations understand that continuous improvement is a long-term journey that will create new jobs, jump-start new product lines, increase wages, and develop a strong workforce.

Another important attribute of lean manufacturing is the benefit of flexible operators, line leads, and line technicians. Flexibility is the key to working effectively on a lean shop floor. Operators can no longer become comfortable performing only one operation or working in only one area. Lean organizations need to cross-train operators and supervisors to allow people to be moved or shifted to accommodate seasonal production levels and absenteeism. Because flexible employees add value to the organization, an incentive program should be in place to encourage continuous learning. A cross-trained, flexible workforce helps a company maintain efficient performance levels on the shop floor. This book provides a guideline for preparing a pay-for-skill program that will encourage your people to learn additional operations and become more flexible and valuable to the organization.

Managing in a lean manufacturing environment is tough, especially in the beginning stages. It can be filled with heartache and pain. The people element is vital, and there is no perfect template to go by when you're dealing with the various attitudes and personalities you will encounter. I discuss the typical mistakes managers make when preparing for lean implementation and offer ways to avoid them.

This book is based on years of lean and kaizen experience—what I witnessed during lean implementation struggles and how those struggles were overcome. To finally realize the positive results you have been promised, your lean program requires strong management commitment, the use of kaizen events, accurate early planning, and a comprehensive lean training program backed by constant monitoring and accountability. It's my hope that this book will become a powerful mentoring guide to focus your efforts in the right direction and ensure that your lean manufacturing journey is successful.

Case Study: How Lean Failed

Adopting lean manufacturing is a journey. Sometimes, even the terminology is confusing, including a phrase in the title of this book: *lean implementation*. The best way to clarify is to say that a lean journey has multiple lean implementations that take place throughout an organization. But no matter how we phrase it, lean implementations can fail, bringing the journey to a halt. Throughout this book I discuss a variety of challenges and common failures in implementing lean manufacturing. I hope my experience will help you get a firmer grasp on what not to do and also will provide guidance for your lean journey.

My career has taken me all over the United States, and I have had the opportunity to meet many professional people and to work with numerous organizations. Each of them experienced successes, as well as challenges, when embarking upon their lean journeys. This chapter tells the story of a heating and air conditioning company in the Southeastern United States that I will call X-Corp.

This company began its lean journey with a lot of enthusiasm. It hired multiple consultants and trainers and scheduled and conducted numerous events, investing much time and money in the effort. This story isn't an attempt to downplay X-Corp's hard work, its commitment to its employees, or its contributions to its community. It is simply a great example of an organization that struggled to implement lean manufacturing but did not experience the desired success.

Gung Ho!

Before starting Kaizen Assembly, my lean manufacturing training and consulting company, I held a variety of positions in the lean field: lean engineer, kaizen coordinator, and corporate lean champion, to name a few. I was involved with lean manufacturing for approximately four years when I was hired at X-Corp as a senior lean manufacturing engineer.

X-Corp embarked on its lean journey in early 2002. I came on board as a permanent employee in January, intending to be one of the organization's main drivers of lean manufacturing. The company had a new plant manager who had a strong desire to implement lean on the production floor. Having a strong background in continuous improvement, naturally I was delighted at the prospect of working with him. He scheduled a meeting for late February to discuss the schedule for the year with upper management. I was invited to the meeting because I was the lead lean engineer on the upcoming projects. X-Corp had also hired a lean consulting firm to help with the journey.

Most of the managers in the room were not aware of lean and appeared to be confused about the goals that were being presented. The plant manager explained that the consulting firm would send its top advisers to the plant during the weeks we were to have kaizen events (improvement projects). I believe that X-Corp paid the firm approximately $12,000 per kaizen event. In addition, X-Corp planned to invest in a series of training sessions that taught value stream mapping, time studies, and visual management (covered in a moment). The training was to be completed before the first kaizen event. With these measures in place, it appeared that X-Corp was moving in the right direction: setting the vision, training the workers, and then implementing. The process was familiar to me, and I totally approved.

Lean Training

After a series of meetings with upper management, everyone became acclimated to the concepts of lean manufacturing and understood why lean is so important. So far, the journey was very positive and rewarding. The plant manager, through his personal commitment to the vision, had generated a sense of passion in upper management, a critical factor in these early stages. With everyone committed, the next item on the agenda was training.

Visual management was the first training to take place. **Visual management** is a philosophy that outlines the importance of making a work area more organized and less cluttered visually so that workers can see problems more quickly and react sooner to rectify them. The company had selected two managers to attend the workshop. That was a mistake. When you train employees on the concepts of lean manufacturing, it is important to involve many people, from various departments. In this way, the trained individuals can return to their respective departments and train additional employees. These trained employees will then be able to lead your lean improvement efforts.

The plant manager sent the production manager, the engineering manager, and me to the visual management workshop. After the training, we returned to the plant, excited about all we had learned. The workshop focused on 5S and the visual workplace. **5S,** the key foundational lean practice, is a philosophy of organization and cleanliness that should be implemented across a company. The five S's are sort, straighten, scrub, standardize, and sustain. Although the trainer never used the term 5S, I knew exactly what she was emphasizing. I realized that 5S was not only needed on the factory floor but also had to be one of the very first initiatives.

The next round of training focused on **value stream mapping,** an extremely useful tool. In this practice, a team creates a map of the current state of a specific process or area; eventually these **current state maps** are used as part of a kaizen event to implement lean improvements. The consulting company hired by X-Corp conducted this training, as well as the reinforcement activities. In contrast with the visual management training, this time X-Corp filled the room with every manager, engineer, and line lead available.

Everyone was quite engaged in the training, and multiple teams were dispatched across the plant to perform the exercises. Once again, everything seemed to be moving in the right direction from a project management perspective.

When the two-day on-site workshop concluded, each team presented its findings, showing areas of waste that needed to be removed or significantly reduced. There was plenty of opportunity for improvement. Most of the current state maps identified issues with inventory, supplier lead times, overproduction, and excessive wait times in the stockroom. The production lines also had ample room for improvement, but the biggest opportunity was outside of manufacturing. Even so, a decision was made to redesign the assembly lines first and postpone improvements in order processing and inventory.

Based on this decision, X-Corp shifted gears, scheduling an on-site time study as well as data collection training, which I considered to be the nuts and bolts training for the journey. Although I was not completely disappointed with the company's decision and change of focus, I would have preferred to see some effort devoted to inventory issues. It would take time to resolve the inventory problems and supplier issues, but I thought one team could have started working on that area. No matter how much X-Corp refined its production lines, if there was an insufficient quantity of good-quality parts, operations could shut down. Although I felt strongly about this, I embraced the decision and maintained my passion for the journey as we proceeded to the task of data collection.

Data collection training went very well. X-Corp registered a lot of employees for the on-site training. The instructor taught everyone how to perform time and motion studies and how to identify waste-removal opportunities from the collected data. Having conducted hundreds of time studies before my employment with X-Corp, I knew how critical the information was for implementing lean manufacturing and removing waste.

The training class was divided into six groups. After one-half day of training, the groups were sent to the production floor to conduct the studies. X-Corp had chosen to send the same people who had participated in the value stream training—a wise decision. Each team was assigned to the process for which it had created current state maps.

I was assigned as the leader for the engine line team. There were four models of engines, each distinguished by a handful of options. The engine, hoses, valves, clamps, and other main components of the engine were all standard. The team gathered the necessary assembly information for each of the four models.

The assembly line was quite long and allowed the buildup of **work in process** (WIP). It was obvious that single piece flow should be implemented to resolve this issue. In **single piece flow,** products are built or assembled one at a time. A single unit is allocated per workstation, with no units allowed to accumulate between workstations. The team also noted that the line operators often left their workstations for non-work-related activities such as taking extra breaks, chatting with passersby, checking personal cell phones, and so on. The team recognized that there were many opportunities for improvement.

X-Corp appeared to be headed on the right path. The plant manager's demonstration of passion and commitment had set the right tone for the journey. The organization was now trained on three major lean topics: visual management, value stream mapping, and data collection. At this point in the game, we were ready for our first kaizen event.

The First Kaizen Event

At this point in X-Corp's lean journey, things were moving along fairly well. We would have benefited from additional training on other lean topics, but, after only three months, the company had made significant progress and team members were still passionate about the journey. But I knew that many companies begin down this path with a lot of zeal, which tends to start fading a few months into the process.

The first kaizen event was scheduled for mid-April, and my excitement was mounting. The lean consultants were scheduled to be available during the week-long event. When a company hires a lean consultant or firm, it is critical to identify its expected level of involvement. When I later became a consultant, I made it my responsibility to guide my clients in the right direction, even if it means rolling up my sleeves and diving in with them. Unfortunately, some consultants are not as involved in the actual process, and many do not provide the level of leadership or direction needed during a kaizen event. Sadly, this was the case at X-Corp. It was during the very first kaizen event that I felt the path for success begin to change direction.

The consultants made a big mistake: They never gave us formal training on kaizen or on how kaizen events are conducted. Our expectation was that we would take the information gained from the value stream mapping and data collection exercises to initiate improvements. Four teams were established, again consisting of the same employees who had participated in the prior training and exercises. Then, as odd as it may seem, the consulting company instructed X-Corp to discard any information that we had obtained. The consultants wanted the four new teams to collect their own information on the first day of the event and then use the data to generate ideas for the rest of the week.

This is traditional kaizen consulting, in which most of the planning is done during the first day of the event. However, in my professional experience, I have found significant value in performing a current state analysis before the event (see Chapter 4), and I highly recommend using this process rather than the traditional approach. The new teams can always find additional ways to improve the process. In the case of X-Corp, the change in direction caused some irritation for those who had performed the earlier analysis, and many felt a bit soured on the whole process.

In an attempt to lift their spirits, I brought my team members together one week before the kaizen event and explained that we would be using the information previously collected. Although everyone was still excited, the group, as a whole, seemed to have lost some energy.

The first day of the kaizen event arrived, and the teams met with the lean consultant to prepare for the week. Except for me, no one had any idea what to do or what was expected. The consultant gave a short inspirational speech on lean manufacturing, and then we were excused to begin our projects. As the team members slowly left the training room, it was apparent that they were confused and unsure of the next action.

I quickly assembled my team and explained that we needed to assess the work content we had captured in the time and motion studies. This information would help us balance the work evenly among the workstations, allowing us to identify what was needed to perform the work. Basically, we were going to sort the area and discard all the unnecessary items from the engine line. This made sense to the team, so we quickly headed for the engine line.

After calculating **takt** (the cycle times for the workstations; discussed in detail in Chapter 5), I divided the team into two groups. We found an area in the plant where we could place the unneeded items. One group remained in that area. The other group took responsibility for identifying unneeded items and removing them from the engine line area. Fortunately, we had two line operators in this group to assist us. The group in the removal placement area simply received the items and then organized them. This system worked very well.

The other kaizen teams, however, were struggling, because they had no direction and no idea what they were supposed to do. The consultant simply walked around and made useless comments to the teams, usually pointing them in a direction that was not correct. Each team had current state data from their respective areas but did not use it because they had been told to ignore it. Clearly, this was a bad beginning for X-Corp's journey and not good for the first kaizen event.

The teams struggled to understand why they were not allowed to use the up-front planning information they had gathered. Over the years, experience has shown me that up-front planning is important. Therefore, I have ignored the traditional kaizen approaches and developed my own system. The traditional approach allows the teams to come up with improvement ideas, on their own, during the kaizen event. In my opinion, you can still get this benefit even after using current state data on the respective process.

After the first day, my team had made significant progress. We had successfully removed all the unnecessary items from our assigned area except the old conveyor. While we waited for the line workers to finish work for the day, I took the opportunity to visit the other teams and my colleagues in the other areas of the factory. They were confused, and the consultant was giving them no direction. Actually, he had left to have dinner with the plant managers and was not planning to return until the next morning. The team leaders assembled for a talk. I explained that our team was using the data collected during the last few months and was moving ahead. The other team leaders saw that this was a good approach and agreed to begin using their information. With management out of the picture for the time being, all the teams started crunching numbers and coming up with line layouts. They worked until about 9:00 p.m.

As the second day began, most of the teams were unhappy with the guidance they had received from the consultants and upper management. It was the consensus that if we had not been told to discard our data, we would be less tired and farther along in the kaizen event. Nevertheless, the clock was ticking, so the teams kept moving.

The teams worked mostly on their own, with a few breaks for meaningless meetings with the consultant. During these meetings, the consultant asked for status reports and gave advice on what the next steps should be. When the consultant offered a smart suggestion, it was either one that had already been implemented by the teams or one that had been mentioned by an employee. Members of upper management, who had not paid attention when a suggestion was originally given, now saw the suggestion as valuable because it had been offered by the consultant. Funny how that works.

With two days completed, the teams were progressing much more smoothly, but, even so, many of the employees were unhappy with this kaizen event. Everyone pushed ahead.

Days 3 and 4 found us facing some of the same challenges. The consultant continually redirected the teams with useless advice. It sometimes appeared that he was trying to prove his worth and justify his price. But each kaizen team had solid team members, and they were all making good decisions on their own. Many of them relied on my input, because I was experienced in kaizen events and believed in a different approach to implementation.

Every night, upper management and the consultant left for dinner at approximately 4:30 p.m., and the teams stayed late, continuing to work. Working late at night is another older kaizen approach, one that is quickly losing its appeal. The old approach requires long days and evenings because no up-front planning is allowed. In contrast, preparing for kaizen events in advance leaves plenty for the team to do during the event but allows for a smoother implementation and avoids long working hours. Working the teams sixteen hours a day during kaizen events does not instill enthusiasm and excitement, nor does it bring about dramatic change in the company culture.

When Thursday arrived, the teams were finishing up their areas. Each team had made improvements to flow, workstation design, inventory quantities, 5S, and standard work. The teams were asked to arrive early on Friday morning to begin assembling their reports. It was a mad scramble. The consultant wanted all the presentations ready by 9:00 a.m. so that he could catch an early flight home.

Each team presented its project, and the company was happy. As always, there were some unfinished items, but the event was considered a success. Directly following the presentations, the plant manager announced that the consultant wanted to tour all the areas. This was a surprise to the teams. The team members ran off to clean up their areas, and chaos ensued. The tour was very fast. The consultant quickly walked through each area and then prepared for his departure. He mentioned how hard everyone had worked and wished us luck in preparing for the next event, which would be sometime in June.

I was frustrated with X-Corp's first approach to a kaizen event. From my perspective, it appeared that the company was simply trying to impress the consultant rather than use him for advice.

Most of the team members, too, were unhappy with the consultant, the company, and their bosses. They were tired and simply wanted to go home for the weekend. They were relieved it was over.

Struggling with Change

Things returned to normal after the kaizen event. Employees returned to their regular jobs after wrapping up loose ends. At this point, I had drawn a few conclusions. First, hiring a lean consultant can be very helpful for your lean journey, but only if you select one who is knowledgeable and proficient, provides hands-on consulting, and has excellent communication skills. The kaizen teams from X-Corp's first event were in constant confusion, and for that I fault the company managers as well as the consultant. They provided either no direction or bad direction.

Second, the company was more concerned about impressing the consultant than listening to its people. Because of the training that everyone had received, each team had solid performers who generally knew what needed to be done. Third, it was apparent that we needed a comprehensive company kaizen program (as explained in Chapter 3). Unfortunately, on this journey, communication, scheduling, team member selection, and up-front planning were minimal. And fourth, X-Corp did not establish goals for the teams or metrics for measuring progress. Chapter 2 describes the strategic purpose of lean manufacturing and explains why it is critical to establish key shop floor metrics. X-Corp had no strategy in place for this, or, if it did, the kaizen teams knew nothing about it.

A company's first kaizen event can be difficult even with a kaizen program in place, so it is important to point out that X-Corp was trying. The company had simply started poorly, and as the senior lean engineer, I needed to recharge and get back on the horse.

X-Corp had now entered the sustaining period. The engine line had gone through a major change, not only in flow and standard work but also in the physical layout of the line, forcing the operators to work as a team. The old layout had allowed operators to build as much WIP as possible on the long conveyor and then walk away. The kaizen team implemented a mobile assembly process, in which the engines were removed from a tote and placed on customized carts designed specifically for the engine line. The line used only seven carts, which moved single piece flow through the six workstations and left no room for WIP. The seventh cart was usually in queue at the first workstation. As the final workstation completed its work and placed the engine into the main assembly process, the line supervisor brought that last cart to the first workstation and placed it in queue.

After the redesign, the space was much more confined but still large enough for the operators to work, maneuver the carts, and find tools and parts. Typically, the line lead stood around watching the assembly of the engine; therefore, he was assigned as the materials handler. We established a three-hour parts replenishment rotation so that he had ample time to address issues as they arose and circulate the carts as needed. All these changes meant that there were many new procedures and protocols to follow.

The operators firmly resisted the modifications, as did the line lead. With the new layout, flexing was now an absolute necessity in order to keep the line moving. **Flexing** is an industry term that simply means "the movement of workers." It is a movement between workstations as needed to ensure that product flows evenly. It is an automatic response to bottlenecks in flow. This was a hard concept for the workers to comprehend. Management had not trained them on flexing or how to identify bottlenecks in the process. Most of the production workers had received very little hands-on lean training, and this was unfortunate because it would have been directly applicable to their environment. There was no system in place to bring the operators up to speed. There were operators on the kaizen teams, but that was the extent of their participation. Therefore, after the changes were implemented, they fell back on old, established patterns of operating, and this created a multitude of problems on the new line.

It was still common for workers to leave their workstations even though the line was under single piece flow and flexing rules. Operators would also leave to retrieve their own parts, because the line lead had refused to take on this new role. Flexing was virtually nonexistent, and many times an operator would stand in the workstation, waiting. The line lead did not make people accountable, the supervisor was rarely available to advise the line lead, and the production manager was never involved. With no accountability and no actions taken to eliminate the resistance on the floor, output and quality suffered.

The need for culture change is the hardest part of the lean journey, and the plant manager and upper management simply did not address this issue enough. Production people were allowed to do as they wanted, and, ultimately, the line's poor performance was blamed on the kaizen teams. Several meetings were held in which the manufacturing engineer and other kaizen team members expressed their dissatisfaction with how the lines were being managed. They complained that the operators and line leads were not following the work content, were refusing to flex, and were still leaving their workstations.

As the lean engineer, I continually advised management to train the operators more formally. I explained that the workers were given a new process and were expected to follow the rules without any notice or up-front training. The problems that the line faced were a result of management's poor planning as well as its approach to employee resistance. The engine line and other newly developed lean processes were falling apart.

Raising the Bar: 5S Implementation

The next kaizen event was a couple of months away, and our focus was still directed toward the struggling engine line and the other processes having difficulty. Because management had not developed a strategic purpose and had not established any way to measure our progress, it was difficult to gauge the financial burden or level of improvement. There was ample room for positive changes, and I felt strongly that the line workers needed a few more things to help them get up to speed. I decided to assemble a small team to start addressing other issues.

On the first day of the first event, the kaizen team had been able to get only so much done, considering that we had been told to start from scratch. However, when we had made the decision to use the time studies, we were able to balance the work content and establish some standard work. Now I wanted to implement 5S in more detail. The kaizen team hadn't accomplished much beyond some bin labeling and work-station signs. My team consisted of the manufacturing and quality engineers on the line, a total of four people. I placed an order for floor paint, floor tape, labels, bins, and other miscellaneous supplies that would improve the visual effects of the line.

We scheduled the work to be done during a week when the line would be finished by noon on Thursday. After the operators left, the team cleared all the items from the floor. Because the line was entirely on wheels, this was a fast exercise. I highly recommend mobile lines.

The team quickly cleaned the floor in preparation for painting. After the floor dried, we grabbed paint rollers and proceeded to apply the paint. I had ordered a standard color, manufacturing gray. To make the line stand out, we painted the entire area designated for the engine line. The yellow floor tape and other visual markings were easy to see in contrast to the gray floor.

My goal was to finish painting by the end of the afternoon, to allow time for the floor to dry overnight. We would start putting up the visual markings and designations the next day. On Friday, we spent all day placing yellow floor tape to designate anything that went on the floor: totes and bins, parts racks, the path of the carts, garbage cans, and the like. The team labeled everything on the floor with descriptions and quantities (when applicable). The floor designations and labeling took very little time, so we decided to make new labels for the parts bins and reorganize the parts racks. After only one day of work, the line looked very nice and much more organized.

The work performed on Friday did not go unnoticed. On Monday morning, the line operators were the first to acknowledge the improvements. A few production supervisors commented on the appearance and asked when their lines were going to be improved. My manager said that she was happy with the line and hoped it would be sustained.

Improving Work Instructions

The initial kaizen team had established single piece flow, standard work, and 5S. I met with the manufacturing engineer assigned to the engine line and discussed the need for point-of-use work instructions. The current work instructions were in poor condition, and because we had made major changes to the work sequence, new work instructions were needed. The old templates were terrible—too many pages and too many words. These templates were stored in a notebook, which was kept on a shelf near the line. I explained that we needed a fresh approach. My goal was to reduce the size of the documents—to only two to three pages per workstation—by using signs and icons to illustrate processes rather than an excessive amount of words.

Making this drastic change to company work instructions required that I meet with upper management to discuss our intentions and get approval. The management team was thrilled and thought the idea was great. Everyone agreed that we needed new lean documentation, and the managers decided that when the engine line had been completed, new work instruction templates would be designed and would become standard.

The engineer and I worked every day, updating the work instructions. We held several meetings with the operators and line leads, and they appeared to be in favor of the modifications. The old work instructions were so poor that they were seldom used. The new format would be beneficial for everyday use by the operators as well as for training new employees.

The new work instructions were completed in about two weeks. After the appropriate approval process, the work instructions were installed in the workstations on the engine line. We felt that the improvements to 5S and documentation would help change the culture and get the supervisor to instill accountability in his people. Time would tell.

Lack of Accountability

The engine line followed the new procedures for about a week, and then problems started to arise. Although the line was much more organized and the workers had all the tools, parts, and documentation they needed, the line was performing poorly. Operators resisted the implementation of standard work and the requirement to work at a more productive pace. They complained about the expected pace; the old line had allowed them to work at whatever speed they wanted to, as well as take multiple breaks. They became stubborn and did not want to work as a team. What's more, the production supervisors did not enforce the new procedures, allowing the operators to do as they wanted. The new standards made this behavior readily apparent, and decreases in productivity, volume, and quality were easily visible.

The new engine line was designed to produce 65 units a day, which equaled a 6.46-minute takt time. In fact the engine line was averaging about 55 units a day and getting increasingly farther behind. The blame fell on the kaizen team, the manufacturing engineer, and me. As a team, we again presented the facts to our management, clearly showing that the line leads and production supervisors were not holding operators accountable or holding them to volume and productivity standards. Blame always fell on the support staff. Clearly, there were commitment issues with upper and middle management.

In an attempt to break through these culture barriers, I scheduled numerous meetings with the necessary people. Sometimes, these meetings were canceled because of lack of attendance or because participation was unproductive. Participants failed to follow up on their action items. Things were quickly falling apart, and I was becoming worried. In addition, during this shaky transition the plant manager was already discussing the next round of kaizen events. Although I am an advocate of continuous improvement, I felt that we needed to resolve the culture issues on the line so that the company could begin seeing quantifiable results. But upper management decided to stop focusing on the engine line because it wasn't making progress. The managers tried to redirect company efforts toward the next kaizen event, and the engine line was left behind and forgotten.

The Second Kaizen Event

At this point, a lot of time and money had been invested in the lean journey and there had been very little reward. It was now July, nearly three months since the first kaizen event. X-Corp was not moving in the direction I was hoping for. There had been a lot of initial excitement and valuable up-front training, both of which are important. However, even with these factors, the implementations were slow and relatively unsuccessful. Some of the other areas that were targeted in the first event slowly returned to their old way of working and discarded many of the improvements. One reason for the return to the old, inefficient work patterns was that new employees were brought on board without the benefit of lean training.

As I've mentioned, it is critical that all employees be trained on, at minimum, the basics of lean manufacturing, as well as how to act and work in a standardized process that is focused on waste reduction. At X-Corp, new operators were placed on the line as soon as they had completed a brief company orientation. Because the production supervisors and operators were not following the new lean processes, the new employees didn't either. As with many companies going through lean transformation, the culture was not committed to the process.

I still had an interest in the success of the engine line, so I kept a focus on it from a distance. It was difficult to walk away from my own line, but I helped as much as I could. The second event was rapidly approaching, and again X-Corp was bringing in consultants. This time, I was hoping for more participation and guidance. We were informed that the company was sending a different consultant this time around. The X-Corp employees were still disgruntled about the first kaizen event because of the lack of communication and structure. An organized kaizen program (as I describe in Chapter 3) was definitely needed. Nevertheless, the teams were formed and the date approached quickly.

This time around, I scheduled a meeting with the team members who were assigned to me. The event was scheduled for August, so we had some time to analyze the process we were given. This particular line built small heating and air conditioning units for small delivery trucks that transported cold products. It was called the 1065 line. These units were about the size of a 25-inch television. I asked each member to collect data for time studies, inventory analysis, on-hand quantities, and waste analysis. This time I was intent on being prepared and going in fully armed, and no one could persuade me differently. I felt it was time to take this lean journey in a different direction and planned to ask my team to perform the duties needed to implement a lean process and really make it stick.

As with the first kaizen event, we were told not to do any up-front planning so that it could be done by the team during kaizen week. Clearly, this approach had not worked well, so my team went to work behind the scenes. As the weeks passed and the event drew nearer, I spent time on the engine line and the other processes that had been part of the first event. It was obvious that management lacked commitment or accountability in these areas. The 5S procedures on the engine line began to fall apart. Workstations became cluttered, labels and designations were not being followed, and operators were still leaving their workstations. Even with production supervisors standing right there, the new standard operating procedures were not followed. Sometimes, the supervisors would strike up 15- to 20-minute conversations with the operators as the line became backed up with units. As a lean engineer, I was losing patience and did not know where to turn. At the same time, I tried to stay positive and keep the new team focused on data collection for the next event.

The first day of the second kaizen event arrived. X-Corp gathered everyone into a training room to meet with the new consultant and hear about the week. My team was ready. The members had gathered a significant amount of information about our assigned line. We were ready to hit the ground running the minute we were released from the meeting. The new consultant gave a speech that sounded a lot like the one given by the first consultant—about waste, change, and kaizen. X-Corp's plant manager gave the teams his go-ahead, and off we went.

The other teams walked out with clipboards and stopwatches. We walked out with drills and tools. There would still be plenty of ideas and solutions generated by the team during the week, but we were excited about being already armed with data on the current state.

Most of our team meetings were held right on the floor as we mapped the new line design, with review and insight provided by the operators and supervisors. Having them involved in the design phase was critical to our success, because it promoted greater acceptance from the other workers on the line.

As the event ran its course, the team felt certain this process was going to work—perhaps not from a design perspective, but definitely from the perspective of culture change. The team slowly pieced the line together. We reduced the number of workstations from ten to seven by removing wasted walking and work content imbalances and by implementing single piece flow. We approached 5S head-on, leaving nothing unidentified or undesignated.

It was a satisfying and exciting transformation. As the lean engineer, I wanted the entire plant and all the teams to succeed. We had initially anticipated that the engine line would set the standard and be the model line, but management and production supervision had let the line fall back into inefficient routines. The 1065 line gave us another chance for success.

The team members worked well together. They came up with creative ways to construct the workstations and present tools and material. My manager pulled me aside and expressed her satisfaction with our progress. Looking directly at her, I said, "That is what good up-front planning can do for you."

The second kaizen event came to an end at X-Corp, and the 1065 line looked great. Of course, the real test would come on the following Monday, when the line would run its new processes for the first time. On the last day of the event, I allowed my team to make the presentation to the company. We had put a lot of effort into the 1065 line and placed great trust in the operators and production supervisors to make it work. They all appeared to be well equipped to handle the new processes, and I assured them that the team members, although going back to their regular jobs, would be there on Monday to support the effort.

Giving Up on Lean

On Monday the line went "live," and all hands were truly on deck. We were determined to make this work. We monitored the line during the entire week, providing support wherever it was needed. The work content needed to be rebalanced a few times, some tools were not in the right places, and there were minor issues with missing parts in the stations—typical minor issues that arise during the transitional phase after a kaizen event. By the end of the week, it appeared that the line was operating to design. Although the operators had not yet reached the newly established design volume, they were getting closer every day. Everything appeared to be going in a positive direction.

During the following week, I began to develop a larger plan for the company's lean journey. I was just beginning to understand the concepts of a companywide kaizen program, and I wanted to put them into action. The organization needed a governing body to watch over all kaizen events, and I felt that I should be given more authority than the outside consultants when it came to planning the events. I created a four-week timeline for each event, outlining everything that needed to take place ahead of time. It took me some time to put the pieces of this kaizen program together.

Periodically, I checked on the 1065 line to observe its productivity, volume, and quality metrics. The line was doing well, and the operators were following all the new procedures and processes, including the staffing requirements. (X-Corp had a habit of throwing people at a process to ensure that output was generated, not realizing the cost of that approach.) Everything was running smoothly—until changes were announced that would affect every line in the factory.

The production manager began changing the roles of the production supervisors. The 1065 line supervisor, who had been involved in the early phases of our implementations, was assigned to supervise a different line. The 1065 line was to be supervised by someone who had no lean background and had not been trained to manage this type of a process. This individual quickly moved people around, added unneeded people, and forced the operators to ignore single piece flow and create excessive work in process, all in an attempt to increase output. Suddenly, the line supervisor and I found ourselves in a fight to keep the process in control. I was patient at first, explaining to the new supervisor how the line was intended to operate. Although he listened intently, he never followed through with the promise he made to follow our procedures.

For weeks, the former team members and I battled with the supervisor and management. We had numerous meetings about the issue, but we were simply told that we needed to work together. We were losing our control, and the 1065 line was rapidly following the same path of failure that the engine line had followed.

I scheduled a meeting with my manager to discuss the problems and get her support. She told me to let go of the issue, because the production workers were going to run things the way they wanted to. This seemed to be a great time to present my ideas on the new approach to lean implementation. She listened to everything, including my complaints about the engine line and the 1065 line. Then she looked at me, paused, and said, "Well, these ideas didn't work on the engine line." I was speechless. She explained that management was unhappy with the way the kaizen teams and leaders had performed their projects over the past eight months. Management was concerned about plant productivity and did not think that the teams had done a good enough job; again, I was speechless.

This was pure management denial. She did not make a single comment about accountability or commitment. Although the teams had made mistakes, as all kaizen teams do, they had worked hard and were very flexible. However, X-Corp's management had decided to stop holding kaizen events, and there would be no more consultants on-site. Two months later, I resigned.

End of the Journey

What happened to cause the failure of this lean implementation? It may appear that I am picking on management, supervisors, and operators more than any other employees. But change comes from the top of an organization and trickles down through the rest of the company. Change can begin with only a few key people. When they embrace change fully, they can begin to change the culture of the rest of the company. Those who manage the operational processes of an organization must be the ones who drive lean implementation and positively present the lean philosophies so that others will embrace them without resistance. Upper management must show total commitment to the process and must demonstrate its ability to hold people accountable for adhering to the changes.

I could have written a highly detailed book about the struggles at X-Corp, but this chapter is sufficient to show how easily an attempt at lean implementation can fall apart because of poor management commitment. I do not mean to imply that engineers, kaizen champions, and technicians do not make mistakes on a lean journey. They do. However, the key players are the ones who set the vision, and change the company culture, with a firm demonstration of commitment and accountability. That responsibility lies with upper management.

Dealing with any change is difficult, and, unfortunately, there is no perfect template to use as a guide. But as you'll learn in Chapter 2, there are ways to define and develop the kind of insight within your company that will help ease your journey. In addition to having commitment from leadership, you can make your lean program much more successful and fulfilling by avoiding or reducing the kind of small mistakes that companies often make, a topic we turn to next.

Apply Navy Sea Teams

A Talent + A Talent — More output

Mission critical initiatives

t w o

The Change Commitment

For years, lean practitioners have been preaching about the importance of getting management's commitment to change. This necessity is as old as the lean philosophy itself. It's well known that a lack of commitment from management is the one definable reason that lean implementations fail. Yet many company leaders embark on a lean journey with initial zeal, only to allow it to fall apart like every other new idea or program.

Why does this kind of failure occur? It is because change can be difficult. Even though positive transformation can result, changing a paradigm, breaking old habits, and discarding established routines can be tough transitions for anyone, management included. In contrast, maintaining the status quo requires very little effort. Business is good and the customers seem to be happy. Profits and cash flow are good, employee turnover is under control, manufacturing processes are running well, and operational conditions appear stable. It is when things are going well that companies often hesitate to implement something new; they don't want to risk what they feel is already a good thing.

If you think you may be perceiving your organization in this way, you must know that you have a large group of supporters out there, supporters who want you to maintain your established methods of operating and eliminate any thought of continuous improvement initiatives: your competitors. They want you to conduct business in the same way you are now. It offers them a great opportunity to gather more of your market share, so they are behind you 100 percent.

Global competition is a fierce reality of the business world. When asked why they chose to embark on a lean journey, my clients commonly respond, "To stay competitive." The principles of lean manufacturing can truly help your business continue to play competitively by operating within a framework of continuous improvement. Do you know why this is important? Your competitors do.

The Three Main Drivers of Product Success

The focus of any competitive business must be its customers and their needs. Potential and existing customers have certain expectations, and manufacturers must try to establish processes that satisfy and exceed those expectations if they wish to remain competitive. In the global economy, people can go almost anywhere to buy products and services, and that is why customer service has become an important factor.

The concept of lean manufacturing has made its way through a variety of competitive industries and has been proven successful. Even so, many companies have not fully embraced the lean philosophy or applied what they have learned about it to their daily operations. And yet these same businesses may wonder why their performance is in jeopardy as compared to their competitors'. Athletic teams and athletes operate under the principle of continuous improvement, constantly seeking to find an advantage, or competitive edge, that will make them more successful as a team; operating a business is no different.

Three main business decision drivers can make or break a deal. When customers come to you, seeking your product or service, they base their decision on cost, quality, and delivery. It is important to find an efficient balance between these drivers, focusing on all three equally, so as not to give one all the attention and allow the others to falter. However, finding an optimal balance of focus can be difficult, and each company defines it differently. The only way to ensure balance is to employ the methodology of continuous improvement, which is the fundamental principle of lean manufacturing.

Lean manufacturing principles, when correctly applied to your environment, can be a powerful tool to determine optimal cost, quality, and delivery for your products. Although there may be a few businesses that always look for the cheapest method, the ideal approach is to find balance. In the following sections, I break down each of the drivers and note how they affect one another within a manufacturing environment.

Cost

[handwritten annotations: OS1, bos develo ch Apple i a 4 l, 1.2,000 MS engand 3 ybs lo vista]

Typically, companies handle cost in one of two ways: cost cutting or cost managing. Those that favor cost cutting employ downsizing, firing, improvising, abuse of suppliers, and cutting corners, all of which are clear cost cutting actions. Some companies even believe that lean manufacturing is cost cutting, but that is simply not true. Rather, lean manufacturing is about managing costs, which makes much more business sense in the long run.

 If you focus only on cost, the quality and potential delivery of your product or service will suffer. In an attempt to lower the cost of labor, many manufacturers try to limit the number of hourly workers and run "thin" assembly lines. Both of these measures constitute cutting costs, but they are usually futile efforts. Typically, the remaining workers are required to pick up the slack and work faster to meet critical deadlines. Essentially, the process is working beyond capacity. This practice is dangerous, because critical assembly and quality steps may be skipped. What typically occurs is that the delivery date cannot be met, and the product is either shipped late or with substandard quality, displeasing the customer and tarnishing the company's image.

As a kaizen practitioner and trainer, I advocate making process improvements without a lot of cost; however, you must have certain necessary items on hand to ensure an effective process. Assembly lines and other manufacturing processes require tools, workbenches, conveyors, lighting, parts presentation, documentation, shelving, bins, tool holders, equipment, fixtures, and jigs, and these items need to be working properly. At some point when preventive maintenance efforts are exhausted, companies must replace items. Line workers cannot perform efficiently using unreliable or broken tools or equipment.

The costs associated with improvements in training and mentoring are sometimes considered too high. That is not smart thinking. In contrast, I recall a situation in which I negotiated a consulting contract with a company located in Nebraska. We were discussing our potential agreement when the owner made it clear that the initial training was something he could not afford *not* to do. As far as he was concerned, training was an absolute must in his ever-changing market. Needless to say, this company is prospering in an industry where companies that are not embracing change are struggling. The owner is beating the competition and meeting the ever-changing needs of his customers. Don't cut your costs. Manage them.

Quality

Forced to choose only one driver to keep at an optimal state, I would choose quality, hands down. By no means should cost or delivery be allowed to falter because of a focus on quality, but customers are more loyal to quality than to any other driver.

We are all consumers seeking products and services. If the quality of the product meets or exceeds our expectations, we will sacrifice a bit on the cost or delivery. After all, it may be worth a little delay or additional cost to know that we can rely on a particular business to provide high-quality products and services. I know that many consumers simply look for the rock-bottom price, but I still find that most people value quality most.

Let me give you a personal example. Like most Americans, I own a car. Owning a car carries a variety of associated costs, such as gasoline and insurance. And if I want my vehicle to last, I must spend money for preventive maintenance. As a consumer and automobile owner, I can take my car anywhere for repairs, and each shop offers some level of cost, quality, and delivery.

There is a quick lube shop down the street from my house, probably five minutes away. This shop offers a multitude of services and products at a low price (cost). The workers are very fast and can have my automobile done in a timely manner (delivery). The problem with this particular shop, which is not representative of other quick lube shops, is that its quality is not to my satisfaction. The workers tend to skip steps, forget services, and make errors in the invoice. This shop also makes a practice of hiring high school kids with limited knowledge of automotive repair. My point is that the quality is not good. Although this shop appears to be balancing the other two drivers quite well, quality suffers, and I am loyal to quality.

So what is my alternative? I choose to take my car to a repair shop farther away from home. This shop tends to take a longer time to get the work done; often, I have to wait in the customer lounge, reading magazines. Its prices are higher than most shops in town, too—but its quality is outstanding. This shop has outstanding customer service, and I never have an issue with regard to its quality of work or documentation. If cost and delivery were more out of balance, I might be tempted to seek out another vendor. That is a real possibility. But for now I am satisfied to pay more and wait longer in exchange for the high quality of the work.

That being said, it is possible to place too much focus on quality, causing cost and delivery to suffer. Manufacturers that focus most of their attention on quality may not trust their ability to manufacture a high-quality product. Perhaps there has been a history of poor quality and they are afraid of losing customers.

Companies that spend a lot of time on quality have a greater chance of increasing operating costs. Assigning more people to focus on quality is expensive, and at some point it becomes redundant. Inspections and testing are not considered value-added activities unless the customer is willing to absorb the cost. Some manufacturers have government contracts that demand strict adherence to specific criteria, something the buyer is willing to pay extra for. In this case, it may be cost-effective to perform repetitive inspections and testing. But most manufacturers do not fall into this category. Delivery is also negatively affected by an overemphasis on quality checks, with extensive lead times because of the extra time needed for checking and rechecking product.

Delivery

"Build, build, build" seems to be the mantra of every manufacturing company. In my ten years as a lean practitioner, I have not yet come across a manufacturer that was not output driven. Of course, deadlines and on-time delivery are not bad things. But speed for its own sake seems to be the cause of much unnecessary chaos. Companies that focus on delivery exclusively will quickly see an increase in cost and a decrease in quality, especially in uncontrolled working environments. Companies that have this volume-driven mind-set cannot see any further than the next hour; there is no long-term vision for the day, week, or month. Crisis management is the name of this game.

The first indicator of an inefficient line layout is a production supervisor or manager who throws people at the process to meet daily production requirements. This practice costs money and does not add inherent value to the flow of the product. The cost of labor simply increases, as does the risk of potential quality problems such as scrap, rework, and future customer complaints. It is hard for many managers and line operators to break this habit, even after lean principles have been applied, but it is a mentality that must be reversed.

The added cost is not usually passed on to the first customer, but eventually it may be added to overall business costs and that will affect future customers. It is a cost that must be absorbed initially by the manufacturer, and therefore it may be pulled from other areas of the company, causing a shortage of funds there. Business 101 is to manage your organization to accommodate the original price agreed on with the client while being able to fund operating costs. Therefore, you are shooting yourself in the foot by creating added costs outside your business structure, and if you compound these costs for each line in the plant for one year, it begins to add up. Additionally, you incur costs that arise from customer complaints, warranty work, and service calls.

Simply put, your organization realizes no gain if you simply create problems and add cost by driving processes and people to meet unrealistic delivery dates. Manufacturing processes should be designed to meet delivery dates efficiently and should be monitored continuously to meet that objective.

In short, the key to managing cost, delivery, and quality is to achieve optimal balance. It's an ongoing challenge. Your first step in getting there is to recognize that customers can go anywhere to do business, and their choice is based on cost, quality, and delivery. Lean manufacturing provides a variety of tools and principles that can help put you in a direction to better find this balance.

The Big Picture

Each business is composed of a web of critical components, all contributing to making the business successful. This web, called a company's **value stream,** includes all the activities that a company, its suppliers, its employees, and its owners are involved in to design, order, produce, and deliver products and services.

Everyone in a value stream has a stake in whether a customer chooses to do business with the company or with a competitor. In a global economy that is continuously expanding, the purchasing options are endless. This is the big picture, as shown in Figure 2.1. Suppliers, employees, and owners all contribute to the success or failure of the organization. Therefore, all should strive to operate in ways that improve cost, quality, and delivery.

Let's look at each of the players in detail.

Owners

The biggest stakeholder is the business owner. Often, the owner (with or without partners) started the business and has a vision of its operation and its future growth. Money and time have been invested. The owners need to think of their families' welfare as well as the welfare of their employees, so job security is a concern. Business owners want strong financial gain, as well as optimum growth that will sustain business profitability. With regard to cost, quality, and delivery, owners are at the forefront in attempting to find this balance.

Employees

Job security is very important to most employees. With length of employment at any one company much shorter than it used to be, employees count on steady paychecks, medical insurance, and investment opportunities, and they also desire an enjoyable and stimulating work environment. Employees have a strong interest in the company achieving an ideal balance between cost, quality, and delivery, because it secures their employment. Employees want customers to choose the products and services provided by their organization. Employees want to contribute to the company's financial strength and continued growth, just as business owners do, but employees' primary motivation is reliable job security.

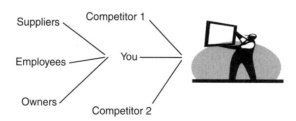

Figure 2.1 The Big Picture

Suppliers

Although business owners may have issues with suppliers from time to time, suppliers play a critical part in obtaining optimum cost, quality, and delivery. Remember that lean manufacturing is not about strong-arming your suppliers but about developing long-term relationships, which involves open communication, mutually agreeable contracts, and the desire for both to achieve success. Suppliers want potential customers to buy your products because it brings them business. By using lean manufacturing tools, many world-class organizations have involved their suppliers in the lean implementation process. As part of the value stream, suppliers need to be a part of streamlined processes and efficient transactions, helping them do business effectively with your company. Even if processes are refined and waste has been significantly reduced, it is all a moot point if suppliers cannot meet the current quality and delivery expectations. When the value stream is improved, everyone is a winner.

Acknowledging and understanding the bigger picture are essential to the success of your lean journey. Everyone in your value stream can contribute to the cause.

The Strategic Purpose

When it comes down to it, commitment to change is essentially a philosophy. Change is not easy for some individuals. In truth, for some people, change is extremely difficult, because it rocks the boat and makes things that were familiar suddenly become unfamiliar.

It is also difficult to get people to embrace change as a positive measure. And there is no perfect template to follow when adapting to change or teaching others to adapt. The challenge lies in getting everyone to see change as valuable and necessary and ultimately the best way to ensure success. Effectively, change is not in our words; it is our actions.

To avoid lean implementation failures on the shop floor, management must first establish the foundation for success by developing goals and metrics to improve cost, quality, and delivery. Improvements in these areas will have a profound impact on the company's financial strength as well as overall growth. Actions should always follow words. Don't simply state, "We are doing lean"; demonstrate it by taking action. This important concept is what I call the strategic purpose.

The **strategic purpose** is an effective way for management to demonstrate its commitment to the lean program—and to the lean philosophy of continuous change and continuous improvement—beyond inspirational speeches and company declarations. A company's strategic purpose serves as a guideline for implementing lean processes and sustaining positive change. (Some individuals refer to the strategic purpose as a **lean strategy.**)

A key part of creating your strategic purpose is to establish a list of critical shop floor metrics that can be measured and quantified. On the production floor, these metrics are often called KPIs, or **key performance indicators.**

Here are the most commonly used shop floor metrics:

- Productivity

- Quality

- Inventory or work in process (WIP)

- Floor space use

- Throughput time

Productivity

Productivity can be measured in a variety of ways. A productivity measurement requires some kind of input: labor dollars per unit, the distance product travels per person, pound per machine, bag per person, and so on. Or you can compare something like labor hours to standard cost hours. All these are examples of productivity measurements.

Productivity is improved when products are manufactured with, for example, less effort, fewer workers, less equipment, and less use of utilities (overhead). Because lean is about managing costs, and not cutting costs, manufacturers need to adopt a smart approach in attempting to achieve minimum effort. Supervisors and managers always seem to be concerned about speed, but productivity is not about speed; rather, it is about **pace.** Over the years, manufacturing professionals have had a misconception about productivity. Working hard and fast to excess is not conducive to good quality and safety. Human beings can sustain 100 percent speed only to a point without negatively affecting quality or seriously negating safety factors. So 100 percent pace makes no sense.

Lean manufacturing is about working smart, at a pace that is sustainable, while safely producing the required number of good-quality products in a given time period. There is a methodical approach to designing a process that creates a smart pace, allowing managers and engineers to calculate accurate worker requirements (I discuss this in Chapter 4, Early Stumbling Blocks).

Productivity is directly related to cost. Therefore, it is the most important shop floor metric in your strategic purpose. Focusing on the efficiency of the production operators is critical, because they are considered value-added; they build product, and that, in turn, pays the company's bills. If you adopt a more efficient process and require fewer personnel, you will dramatically improve cost. Using fewer personnel equates to, for example, less labor, fewer tools, fewer workstations, less documentation, and less material. Fewer people handling product also equates to improved quality. Unnecessary people working on the product, in an uncontrolled and poorly designed process, can increase the chance of quality errors, not to mention safety issues. It also improves delivery because it means having fewer processes, fewer people, fewer transactions and systems, and fewer places the product must travel before landing in the customer's hands. These are key points that contradict the old philosophy of simply throwing more bodies at a line in order to meet deadlines.

The important thing to remember is that it's best to manage costs and redeploy personnel as areas become more proficient due to improved processes. Don't just cut people! That is not the lean philosophy. Proper worker allocation is the best approach.

Quality

Quality is one of the key shop floor metrics in the strategic purpose. However, before adding it to the list of metrics, you must make a decision: how you will measure quality. Surprisingly, I have witnessed many companies, large and small, that do not measure quality—a big mistake. Quality should be measured both internally and externally. Some companies use customer complaints as the external measure, simply keeping track of the number of complaints per month and the cost of problem resolution—warranty costs, cost of service calls, and so on. Although those costs are important, you should also track internal costs: rework costs, scrap costs, parts per million (PPM), the number of rejections, and so on.

Quality can be measured in a variety of ways, and it depends on your products and processes. No matter which method you choose, a measurement of some kind must be in place.

It is difficult to improve quality if the production process has significant design variability, a lack of cross-training among line operators, unreliable equipment, outdated documentation, or poor flow. Although it takes many steps to refine these elements, it is difficult to gauge quality improvements until refinement occurs (as discussed in Chapter 4). Quality is improved when you reduce the number of defects, rework, scrap, and external complaints. With this error reduction, you will spend less time and money reacting to problems after they occur. Delivery will be improved when you reduce downtime created by line stoppages. A controlled manufacturing environment promotes good quality, with predictable lead times, and these factors will definitely please customers.

Inventory and WIP

A lot of money is tied up in parts and material, and that is why they should be part of your metrics, providing a method for monitoring and improvement. However, reduction of inventory to a satisfactory level cannot be done overnight. This metric also includes work in process (WIP): parts, subassemblies, partially completed units, and finished goods. The ideal state occurs when completed units have been loaded onto a delivery truck and are not just sitting around in a physical or logical location known as **finished goods inventory.** Unless your plant is used as a distribution center, fully built products should leave your facility quickly.

Many manufacturing facilities literally use their production lines as stockrooms, with large quantities of parts and material lying around in the assembly area. Some managers view this as a positive method of reducing indirect labor costs because fewer workers are needed to bring material to the work area. Although this practice might make sense in the short term, you incur a much higher cost by ordering large quantities of parts and then storing them. By allocating a smaller, controlled number of parts and material to the process, a company can afford to have personnel to manage it. This cost is less than the cost of excessive inventory.

Excessive inventory and WIP also create other problems. First, there are storage and handling costs. Large lots of parts or WIP can create quality problems due to overstacking, forklift damage, and hidden product defects that are undetectable in large lots. A huge infrastructure is required to control and monitor excessive parts and WIP. Inventory takes up valuable space that could be used for more profitability, such as building a new product. I may be painting an ideal picture, but it is important that each organization find a healthy balance between parts quantity and the cost of indirect labor (addressed in more detail in Chapter 4).

Using inventory as a metric in the strategic purpose is a critical part of achieving good cost, quality, and delivery. Cost is significantly reduced if you do not have money tied up in excessive inventory or WIP, which is a burden on business operations. Quality is significantly improved, damage is less likely to occur, and initial defects can be better visualized in smaller lots. Delivery is also improved because time is not spent moving large lots from one location to another or managing finished goods that could already have been shipped.

Floor Space Use

Floor space comes at a premium. Renting, leasing, or buying a manufacturing building is one of the highest costs of overhead. Using floor space efficiently is critical to the success of any lean journey, especially with regard to cost, quality, and delivery.

The production floor serves one major purpose: to support the building of products. Although the factory may be used for other purposes (such as holding inventory, shipping, receiving, and maintenance), the production floor should be effectively used for value-added work: building products. Non-value-added work on the floor means less profit, especially considering the high cost of owning, renting, or leasing the building. Over time, items such as workbenches, garbage cans, chairs, machines, tools, tables, carts, parts, and pallets tend to accumulate and valuable production space disappears. Thus, it's critical to include in the strategic purpose metrics for managing floor space. As metrics are monitored and improvements occur, valuable space will become free, allowing you to add more production lines, build more products, and become more profitable.

If vacant space exists without clear visual objectives for use, the space will get filled. That is only human nature. Unnecessary items will begin to pile up and consume production space, either on the floor or in the work areas. In Chapter 4, I discuss the adoption of 5S and the visual workplace and describe common mistakes people make when attempting to reduce floor space.

Cost is significantly improved when floor space is better used. Adding production lines for new products brings in additional revenue. Or if you determine that you need less floor space, you can move to a smaller location, saving money on rent or lease. Quality is significantly improved when less space is used to store unnecessary items that can become damaged or broken. Also, less clutter in the area promotes better visibility throughout the factory. Delivery is significantly improved through the reduction of unnecessarily long assembly lines and other manufacturing processes that could make more efficient use of space. Products and parts spend less time traveling around the factory, and that means shorter assembly lines, shorter processes, and ultimately, shorter lead times for customers. Does it make sense to spend thousands of dollars on monthly lease payments for a factory floor that is only 50 percent used? Floor space is expensive; use it wisely.

Throughput Time

Throughput time is the time it takes a product to flow down the assembly and manufacturing process. Obviously, throughput time has a direct impact on delivery. The more time a product takes to go through the main process, the longer a customer must wait. Of course, there are a multitude of variables that can extend product lead time, so it is wise to simplify the metric by monitoring the throughput time. As an alternative, you can measure travel distance instead of throughput time. It's up to you.

Longer production lines require more workstations, workers, tools, workbenches, conveyors, supplies, parts, and material, and that results in additional costs and WIP as well as extended lead times. I have witnessed some impressive improvements in throughput time; I have seen travel distance reduced from 300 feet to 35 feet, and from 550 feet to 100 feet. A physical reduction in distance equates to less throughput time, allowing an organization to promise competitive, reasonable delivery dates. Also, from a visibility standpoint, it allows a production supervisor to see everything taking place in his or her area from a single vantage point. The less time the product spends in the building, the better. By improving throughput time, you benefit in cost, quality, and delivery.

Improving key shop floor metrics will have a profound impact on the overall financial success and long-term growth of the organization. The metrics discussed here are particularly important because they are directly related to the work being performed on the production floor, and that affects profit and, ultimately, employee paychecks. Production workers need to work in an efficient environment in order to be successful contributors to optimal cost, quality, and delivery.

Creating Your Strategic Purpose

I realize that creating a lean strategy cannot be learned simply by reading a book. There are many variables involved, and you must obtain time commitments from all the departments. However, I can provide you with guidelines for creating this valuable tool, which you can use to navigate throughout your lean journey.

The most important element of the strategic purpose is the key shop floor metrics. Although I have listed the most common metrics, you can develop others for office and administrative functions as well. As you prepare this important document, keep in mind that you can't improve these metrics simply by making improvements to the production floor. Techniques such as new line design, 5S, setup reduction, mistake-proofing, workstation design, single piece flow, and visual management result in an efficient manufacturing process, but other visual parts of the business also contribute to the success of the shop floor.

With that in mind, let's look at writing your strategic purpose document. You'll divide it into two major categories:

- Estimated annual improvement to the metrics
- Departmental responsibilities

Estimated Annual Improvement to the Metrics

You should set annual improvement goals for each of the metrics, indicating the estimated improvement for productivity, quality, inventory and WIP, floor space use, and throughput time. By what percentage does the company want to increase productivity and reduce floor space? What goals would be considered reasonable and obtainable? Managers should all agree that goals should be attainable but challenging, thereby ensuring that substantial improvement is made and justifying the effort.

Productivity

Productivity improves when products are built with minimal effort—for example, less labor, fewer tools, fewer workstations, less documentation, and less material. First, sample data should be collected from the targeted assembly line or process to reveal the current state of operation. Lean manufacturing is all about data—collecting, analyzing, and improving upon it. Improvement decisions are based on the current state, and therefore it's crucial to gather information on the current state of business processes.

There are several ways to collect this data, including time and motion studies, waste analysis, process mapping, and value stream mapping. This data will reveal waste removal opportunities in overproduction, overprocessing, wait time, inventory, defects, transportation, and motion—the so-called seven deadly wastes. Once you've identified a type of waste, you can determine the reason for it as well as the amount of time spent creating it.

After you make these critical measurements, you make logical decisions on methods for removal. Accurate metrics allow you to determine better staffing and workstation requirements for a specific process.

Of course, you may make mistakes when performing waste-reduction metrics, but it is reasonable to assume that overall plant productivity can be improved by 20 to 30 percent over the course of a year. If you practice the suggestions outlined in this book, this rate of improvement is doable.

Quality

After lean processes are implemented, it is easy to see remarkable improvements in quality. Simply implementing 5S and adopting good organizational practices can result in a major decrease in problems related to quality, and that equates to lowered costs. Quality is not inspected into a product; relying on an end-of-line or end-of-process inspection or test does not ensure that quality is built into the product. The responsibility for quality lies with those who fabricate and assemble the parts. Line operators are the manufacturing floor's first line of defense in the fight for outstanding quality. Quality should be designed into the manufacturing process, requiring line operators to check for critical quality attributes throughout the build process. This concept is called **quality at the source.**

A small series of incoming and outgoing checks should become standard for each worker. It is important not to overload line workers with excessive quality checks, but having them check certain aspects of the product throughout the process is critical. Of course, other variables, such as material from suppliers, office errors, training, and machine capabilities, affect the quality of a product. However, the concept of quality at the source can be applied companywide and not only on the production floor.

As the line operators (or any other employees) prepare to perform work, they should always begin by checking the work of the preceding operation. When that is complete, they should perform their own work and then check their work. As these checks are conducted throughout the process, the likelihood of a defect occurring is significantly reduced. Implementation of quality at the source can result in an 80 percent improvement in quality, depending on the metric.

Be aggressive in your goals for quality. A 70 to 90 percent improvement in quality on the shop floor within one year is a good benchmark, depending, of course, on the number of lines and other manufacturing processes in operation.

A machine shop in North Carolina realized a dramatic improvement in overall plant quality. In this case, quality was measured by scrap cost. The organization analyzed its top ten scrap items and then implemented a process in which machine operators and assemblers performed a check on those items, depending on their work content. That one simple improvement decreased scrap costs by 90 percent. That's impressive!

Inventory and Work in Process

Reducing inventory takes time and involves numerous people and departments. It also depends on established relationships with suppliers and other outside vendors, including sister plants. However, you can take several actions that will make a positive impact on inventory kept on the floor.

Reducing workstations and overall line length will help reduce the amount of inventory stored within your manufacturing processes. Fewer workstations equates to fewer parts and WIP. The closer you can get to single piece flow, the better. However, many types of manufacturing industries cannot use single piece flow. The alternative for these businesses is **controlled batches,** in which a certain amount of WIP is allocated in the space between each process or workstation.

Using single piece flow or controlled batches allows you to better estimate the number of parts required for a work area. Avoid storing huge pallets or totes for parts in the work area, because that uses valuable space. For example, if a line must produce 40 units per day and it requires one wire harness, don't store 500 wire harnesses in the work space. Leave that quantity in the stockroom or common storage areas.

You can create annual goals for WIP reduction based on how many lines will be physically shortened. I have witnessed WIP reduced by 70 percent following the successful implementation of single piece flow. Any improvement goal should be based on how many processes will be transformed into leaner, smoother operations. With a full lean implementation, it is possible to realize a 30 to 40 percent reduction in WIP and inventory in a single year.

Floor Space

Manufacturing companies often use more floor space than they really need. Setting a goal for reduction can be a little more aggressive because improvements in floor space are generally greater than improvements in productivity. Keep in mind that all of the metrics discussed in this chapter are intertwined; each affects others. After you perform a comprehensive waste analysis and take steps to reduce waste, floor space will be considerably reduced due to removal of unnecessary workbenches, tools, chairs, conveyors, and so on, and therefore productivity will be significantly improved.

Floor space can also be reduced through the implementation of 5S (sort, straighten, scrub, standardize, and sustain). **Sort** is defined as removing all unnecessary items from the work area, and it is the first step in implementing 5S. Unnecessary items take up valuable floor space that can be used for value-added production space. The goal you create for floor space reduction is contingent on how successful you are in removing all the unneeded items in the production area and also on how well the space is organized and maintained. Data collection and 5S will become two vital components in the early stages of your lean journey. A 30 to 40 percent reduction in floor space use is attainable.

Throughput Time

Long assembly lines breed waste and provide space for items to accumulate. Throughput time is significantly reduced when lines are shorter. Granted, all the necessary people, parts, and tools need to be in place, but the longer the line, the greater the throughput time. Reduction of floor space and reduction of throughput time go hand in hand. There is a systematic approach to waste analysis: get the status of the current state of the work content, and perform time studies. This data is a necessary starting point for making improvements that can eventually shrink the physical length of all manufacturing processes.

It is also important to identify how parts and material are presented to the workstations and the overall line. Storing too many unnecessary parts in an area increases the length of the line. Adding a materials handler to the process increases labor cost, but it reduces line costs overall and positively affects both quality and delivery. The cost of this indirect labor is far less than storing more inventory on the floor, something that results in extended delivery dates. As with floor space, you can set a goal of reducing throughput time 30 to 40 percent for the first year.

Departmental Responsibilities

The shop floor is only one part of the overall lean strategy. Departments such as finance, accounting, customer service, purchasing, engineering, shipping, receiving, and others also contribute to achieving a good balance between cost, quality, and delivery. This approach to the lean enterprise, although a bit unconventional, makes good sense. With a strategic purpose in place and metrics established for the production floor, an organization gains a good sense of why it is in business. Although the shop floor is the key, each department should eventually develop its own internal metrics for improvement.

No single department makes, or breaks, a business. Each department is in place because it contributes to the overall success of the company. Each should be striving for continuous improvement and making its individual processes efficient and free of waste. Improvement initiatives in other departments—such as purchasing, engineering, sales and marketing, production control, and maintenance—should be centered on how they will affect the production floor. Manufacturing companies have a variety of functions that support the manufacturing floor. (Even though some smaller companies may not have all of these specific departments, each service or function is being performed.) Improving the key shop floor metrics should be the ultimate goal of each function, service, or department.

Purchasing

Remember that the goal is to create a balance of cost, quality, and delivery by the improvement of key shop floor metrics. The general function of a purchasing department is to buy parts and material for the production floor and to develop professional relationships with suppliers. Purchasing plays a critical role in a lean organization. No matter how refined, waste free, and streamlined the manufacturing processes are, if quality parts are not available when needed, the line stops. That is not good for delivery or quality.

Establishing annual goals for productivity is difficult when the parts you use are of substandard quality, resulting in production downtime due to confusion, excessive inspection, delay, and rework. These problems hurt productivity. Can your suppliers keep up with the higher volumes you now need because of your ability to reduce waste and inefficiency? Are they able to supply you with parts and material in the quantities and container sizes that will contribute to inventory reduction? It takes time to develop the type of relationships with suppliers that will allow you to work toward mutual goals. But the results will be worth the effort.

Engineering

Engineering plays a major role in shop floor metrics. It is the responsibility of manufacturing and industrial engineers to collect the necessary data for the metrics. Capturing time and motion studies and analyzing waste are their responsibilities. Engineers are responsible for the technical and analytical projects of lean manufacturing and can assist in planning new line layouts, analyzing equipment capabilities, providing input on process quality, and monitoring conformance with new line balancing. Chapter 3 outlines the importance of having a kaizen champion, someone who is 100 percent dedicated to lean and kaizen principles. This individual should come from the engineering department.

Design and product engineers who are submitting engineering change requests (ECRs) should be trained in lean manufacturing and must be part of the lean effort. When parts changes and product updates occur, it affects workstation layout, 5S, line layout, parts presentation, and tools requirements. There must be a methodical approach to ECRs, as well as knowledge of how changes to documentation and other areas can affect the lean process. You need to address work content and cycle times before adding or removing parts from a product line being assembled in a lean environment. Design and manufacturing engineers should be in constant communication. A smart and thorough approach to engineering changes will prevent inventory, floor space, quality, and productivity from being negatively affected.

Sales and Marketing

Does your sales team know your current capacity? Have salespeople been trained in lean manufacturing so that they can make feasible promises to customers? I am often asked why the sales department should be involved in this journey. Lean processes are controlled through procedures, protocols, and rules. Although managers and supervisors sometimes need to make decisions that require deviation from procedures, that is an exception to the rule. Assembly lines and other manufacturing processes are designed based upon data and specified volume requirements. This volume requirement is used to establish the correct number of workstations, people, tools, parts, support mechanisms, and machines. If promises made by a sales team are forcing the production floor to work beyond its controlled limits, it is difficult to balance cost, quality, and delivery.

However, lean manufacturing is about flexibility and your ability to react positively and effectively to changes in customers' needs. What you need to avoid is having an untrained, uninformed sales force that is not aware of the operational constraints of the manufacturing department. Bring your salespeople into the lean process early on; share the lean philosophies and the realities of the operation. Their awareness will help them avoid creating unpleasant scenarios for all company employees as well as for the customer.

Production Control

The production control department represents a major bridge between office functions and the production floor. Often, production cannot move until the official order is received from production control. The customer doesn't care how long it takes for the order to get processed through your organization; the customer cares only about receiving your product by the promised delivery date. The longer an order sits in administrative processes, the longer the customer has to wait for the product.

For small manufacturers, such as a traditional job shop environment where the day-to-day operations are dictated by the number of jobs in the building, the production control department must be efficient. Does the work order sit in a slow-moving stack? How many departments does it have to go through? Who checks it? What are the wait times between checks? In many cases, the production control department is the company's lifeblood. This department can have a significant impact—positive or negative—on productivity and delivery.

Maintenance

No matter how lean a process is, as long as any equipment or machines are unreliable, break down often, or do not operate correctly, delivery and productivity will be compromised. Preventive maintenance is critical to ensure that all mechanical tools on the floor are performing optimally. It is equally important to purchase good-value (not necessarily highest-cost) equipment. Good value means good service as well. Costs must be managed at every level of the organization, including the acquisition of equipment.

Chapter Wrap-Up

Focusing on cost, quality, and delivery is not a new concept. Even in a traditional manufacturing environment, these three main drivers are the cornerstones of doing business. Lean consultants and trainers talk the same language in regard to these drivers, and lean manufacturing provides a lot of the tools to help you need to achieve an optimal balance. It is only smart business.

I recommend that you develop your strategic purpose and key shop floor metrics as a guide to improving cost, quality, and delivery, an effective way to begin any lean journey. Creating the strategic purpose is critical, and much attention should be given to it. It will be a living document that evolves over time as improvements are made and new ideas for improvement are created. Striving for optimal cost, quality, and delivery will be a continuous process.

Although I've discussed only a few departments in this chapter, every-one has a role to play in the lean journey. Each department should create a long list of waste-reduction action items. Support departments such as administration and human resources also need to become more efficient, because how they operate ultimately affects shop floor metrics. That's why the strategic purpose identifies the total vision for the organization, allowing everyone to work toward continuous improvement daily. When the floor is successful, the organization is successful.

three

The Lean Infrastructure: Kaizen

In the early stages of a lean manufacturing journey, companies are faced with a variety of challenges and obstacles that impede how effective and quickly new lean processes are implemented. Engineers, middle managers, and production supervisors are commonly made responsible for integrating lean manufacturing in addition to their regular job obligations. Manufacturing engineers, for example, typically have a collection of time-intensive responsibilities such as writing procedures, updating bills of materials, dealing with day-to-day issues, and even training. Middle managers are typically even busier, because they have authority over multiple people, attend numerous meetings, and participate in decision making. And let's not forget the floor supervisors who run the production lines and processes, interacting with operators and line leads, putting out fires, and managing hour-by-hour crises. When do they all have time to "do lean"?

How can you implement lean manufacturing in a manner that is organized, smart, and effective? Often missing from a company's lean manufacturing program is a firm foundation that embraces continuous improvement. This critical foundation is known as kaizen. **Kaizen** is a Japanese word that means continuous improvement; it is a philosophy driven by the entire company, with each employee involved in improving the organization. The concept of kaizen is not new, but turning this philosophy into action can be difficult.

If you are preparing to embark on a lean manufacturing journey or have started on the path to world-class status, kaizen is an extremely valuable tool to have at your disposal. Knowing that you have the concept of continuous improvement, what's the next step? One of the best methods is to create a companywide kaizen program that is established as the lean infrastructure. This chapter is devoted to the setup of such a kaizen program. The information in this chapter comes from one of my most popular training courses and outlines the important elements in developing and sustaining a program of continuous improvement.

Creating the Company Kaizen Program

To ensure a successful lean manufacturing implementation, you must have specific elements in place:

- Kaizen champion
- Kaizen events
- Kaizen steering committee
- Kaizen event tracking and scheduling
- Kaizen event communication
- Monthly kaizen meeting

Kaizen Champion

In the most successful lean and kaizen programs I have implemented, the organizations designated one individual as the kaizen champion. A **kaizen champion** is an employee who is 100 percent dedicated to the planning, execution, and follow-up of all kaizen events. This person lives and breathes continuous improvement and is constantly moving around the factory, driving lean into the manufacturing processes. More importantly, the champion carries the lean torch throughout the company and has the authority to pull in resources as needed.

The ideal kaizen champion should have a background in both lean manufacturing and project management. Typically, a lean manufacturing engineer is the best candidate, because he or she possesses the skills without additional training. If a lean manufacturing engineer is not available, a manufacturing or industrial engineer can step into the role. The key is to dedicate an individual who is focused on lean manufacturing and will be the catalyst to ignite your lean program and keep kaizen alive.

Selection of a kaizen champion is an important decision that should not be taken lightly. First, you must decide whether you will select a candidate from within your company or hire an improvement guru from the outside. Of course, there are pros and cons to each option, which I discuss in a moment. If you are going to choose an existing employee, he should not come from middle management. Do not appoint an engineering manager, a production manager, or a human resource manager as your kaizen champion. The individual you select must be able to focus full-time on the kaizen program and possess certain skills and abilities.

The candidate must be knowledgeable and skilled in the following areas:

- Kaizen and kaizen events
- Project management
- Teamwork and leadership
- Visual management
- Single piece flow and pull systems
- 5S and the visual workplace
- Workstation, work cell, and assembly line design

Additionally, the candidate must be able to perform the following activities:

- Conduct accurate time and motion studies
- Identify waste and formulate ways for its removal or reduction
- Perform line balancing
- Calculate workstation, equipment, and staffing requirements
- Establish standard work
- Use a computer aided drafting (CAD) software application

Internal employees selected as kaizen champions have an inherent knowledge of line operations and have worked within the current culture that you are preparing to change. An existing employee knows the products that are built and the processes by which they are made. These are important attributes and can carry a lot of weight when you decide on your candidate.

However, selecting someone from within the company can also have a negative impact on your lean journey. Employees who have worked for one organization for a long time have become familiar with the old, established methods as well as any poor habits that may have developed over time. Both support staff and floor personnel can demonstrate resistance to change and become stubborn or hard-nosed in the face of new methodologies. It may be difficult for them to accept different approaches if they have been part of a culture that has not previously been motivated by continuous improvement. Additionally, when an internal employee is assigned to the role of kaizen champion, her regular duties and assignments must be reallocated to other employees. If resources are limited, this could be an issue.

An external candidate brings a fresh perspective to the company. A new employee brings experience and knowledge from working in other factories and can offer a new pair of eyes for viewing change. If the candidate has a background in lean manufacturing, the cost of obtaining training and consulting is significantly reduced. The one limitation of an external kaizen champion is that he is not yet familiar with the company's current product line or how it is manufactured. More important, he has not developed a relationship with the production workers.

Whether hired from the outside or inside, a kaizen champion is a great addition to any company kaizen program and carries much of the responsibility for its overall success. Once appointed, she must remain 100 percent focused on lean manufacturing and must not be deflected from this focus to do unrelated tasks. The key to your success is to allow your kaizen champion to make decisions on her own and to communicate directly with middle management.

Kaizen Events

One of the most effective ways to implement lean manufacturing is by holding kaizen events. Often called **rapid improvement projects,** kaizen events are carried out by a group of management-selected employees tasked with analyzing, designing, and employing lean manufacturing principles within a specific work area. When planned and performed correctly, kaizen events can become an invaluable tool on your lean journey. However, many organizations fail to realize any improvements after conducting kaizen events, usually because of a lack of planning and commitment. Kaizen events require solid up-front planning, a strong team leader, knowledgeable team members, and clearly established goals.

You can notice a lack of planning and improper execution of a kaizen event shortly after the event concludes. To the employees on the production floor, failed kaizen events can quickly become viewed as unnecessary nuisances, and that kind of result is detrimental to the overall implementation. A critical part of planning is to gain management commitment and dedication to lean manufacturing, as discussed in Chapter 2. After commitment has been established, you need to consider several items in order to ensure successful kaizen events:

- Kaizen event selection
- Kaizen team selection
- Kaizen leader selection
- Kaizen event date and length
- Kaizen team goals and expected results
- Kaizen event planning
- Kaizen event budgeting

Kaizen Event Selection

A good rule of thumb is to schedule kaizen events about four weeks ahead of time. This allows for a variety of planning tasks to be completed before the event. When you select the area or process for the event, it is important to consider the effect that the event will have on the company's key shop floor metrics, as described in Chapter 2. To determine whether a particular area warrants priority status, you should analyze its productivity, quality, floor space use, number of workstations, inventory, and travel distance and compare these metrics to those of other processes.

Remember, improving these key shop floor metrics will have a positive effect on company performance, costs, and future growth. Some of my clients have simply opted to choose the product line that had the highest profit margin for their first kaizen event. If you consider profit margin a key business metric, then include it in your measurements. The metrics that I recommend have worked extremely well in the past, and they are quickly traceable after a kaizen event. Nevertheless, you should choose a target based on a metric that will have a profound impact on the financial strength and future growth of the organization.

Kaizen Team Selection

Team member selection is probably the most critical aspect of planning kaizen events. These events are a great vehicle for changing the culture and encouraging people to become more engaged in the philosophy of lean manufacturing. As more kaizen events are scheduled and held, more employees become personally involved in the changes made to the floor, and the culture begins to change. Kaizen event participation should eventually become a requirement, and employee efforts can then be included in performance evaluations.

Successful kaizen events require participation by individuals who bring a much-needed level of expertise and experience; therefore, you should begin by creating a list of criteria for membership. Then, at least four weeks before the scheduled event, make a tentative list of possible team members so that managers can begin preparing for their absence, can verify that vacation requests do not conflict with event timing, and can apply various other employee-related specifics. Then finalize the list two weeks before the event, ensuring that any personal or family restrictions are accommodated, especially if the event will take place outside a regular shift.

Following is a list of potential team members. It is important to have a good collection of disciplines and titles on a kaizen team.

Manufacturing or Industrial Engineer

Depending on the size of your organization, select an engineer who has skills that are comparable to those of a manufacturing or industrial engineer. Many companies assign engineers to particular areas or lines. Over time, they develop a good understanding of the line's products, process, documentation, and operator training levels. For example, if you choose the T100 line for a kaizen event, the engineer assigned to the T100 line should be on the kaizen team.

Quality Engineer

Not all companies have dedicated quality engineers assigned to particular areas. Many quality engineers have responsibilities all over the plant. However, a quality engineer or a highly skilled quality technician should be on every team. As I've mentioned, during the event work content may be shifted, standards will be improved or added, and work instructions will need to be updated. The quality engineer needs to help supervise and oversee these changes and provide feedback on quality issues that may arise from the changes. Material presentation is likely to be altered and improved, and the quality engineer can help with that. Your quality engineer or technician will be able to generate reject, scrap, and other quality metrics from the line and find ways to implement quality standards to reduce those occurrences.

Facilities and Maintenance Personnel

An important member on any kaizen team is a facilities specialist. These people are critical to helping the team finish on schedule and avoid creating delays for the process on the following day. Machines, equipment, workstations, tools, and other elements of an assembly line will be moved around while the line is constructed. Your facilities specialist understands the electrical, plumbing, compressed air, and power systems as well as the networking arrangements and can ensure that all systems are fully operational and all equipment is properly secured to the floor.

Materials Operator

The kaizen team should contain a materials operator on the assembly line specified for the event. A materials operator provides the production operators with the necessary parts to build the products. Therefore, the materials operator has a good understanding of part sizes, and this understanding will be extremely helpful when the team makes decisions regarding materials presentation to the workstations. This individual will also know which parts are heavy or cumbersome and can assist the team in establishing delivery frequencies.

Line Operators

The most important team members on any kaizen team are the operators who work day in and day out on the assembly line. They hold detailed knowledge of the product, the assembly line, the tools and fixtures, the machine capabilities, the daily issues, and other important aspects related to building the products—the kind of knowledge that can be obtained only by actually performing workstation tasks. Two line operators should be included on the kaizen team, and they should be involved in every step of the analysis phase, design, and construction.

The operators will also be critical change agents, working towards revising the culture of the line, and can assist with the initial training of other operators, as well as encouraging sustained change after modifications are made. Other line operators will respect the fact that "one of their own" was part of the team and will be more receptive to the necessary changes.

Management

Top managers should be included on every kaizen team. If possible, plant managers and presidents should roll up their sleeves and get dirty along with the rest of the employees, thereby demonstrating the company's commitment to the success of the kaizen program. Your organization should require members of the kaizen steering committee and other company managers to be actively involved in all kaizen events.

Kaizen Team Leader Selection

Strong leadership is the key to any lean manufacturing program. The kaizen champion is typically the team leader for a company's initial kaizen events. But eventually, each kaizen event should be led by a different individual.

A kaizen team leader should have certain skills and attributes to ensure that he can properly lead people and projects. Each team leader must have an understanding of factory floor lean manufacturing concepts such as waste reduction, 5S, standard work, visual management, and single piece flow. When you assign a team leader, your kaizen champion should always be your first choice, because he has the required background, knowledge, and training. Then, in conjunction with your kaizen champion, you should establish the qualification criteria for additional leaders.

In the past, I have recommended the following characteristics in a team leader:

- Past participation in three kaizen events
- Completion of some type of project management training
- Creation of a budget and schedule for a previous project
- Leadership and work experience in a team environment
- Ability to work well under pressure and within a time schedule

These are basic guidelines, of course, and every organization should establish the specific standards that qualify an employee to be a kaizen team leader. Your kaizen champion will be an integral part of the training and mentoring of all future kaizen team leaders. To promote culture change, I recommend the rotation of kaizen team leaders, allowing new leaders to sign up for the responsibility and become part of the continuous improvement projects.

Kaizen Event Date and Length

I have participated in more than 150 kaizen events as the on-site consultant, a team leader, or a team member, and each one of the events was a unique experience. Traditionally, kaizen events last five days, with the final day reserved for report presentation and a tour of the area. However, the five-day plan does not apply to all kaizen events; some events last only two or three days, and others last two to four weeks. The length of the event depends on the complexity of the line or area and the number of people involved.

Kaizen events can be conducted at any time, including weekends. However, I recommend that the kaizen team not work a 16-hour day, which is a common proposal by many kaizen trainers. Neither am I in favor of having the team work on a third shift, because it can be very unsafe. If the five-day approach is used, the team can work during the first or second shift or over a weekend. Working during the first or second shift allows teams to make changes in real time, while line operators are actually building product. This practice can be a bit difficult, but if good planning and execution take place, it is definitely possible. For companies that operate only one shift, it can be effective to schedule the kaizen event during the second shift, when the line is vacant and tasks can be easily accomplished. Another option is to begin the event on a Wednesday and then work over the following weekend or begin the event on the weekend and work three days into the week. The weekend approach allows more flexibility because the plant is not in operation. The choice is a management decision based on operational conditions.

Planning for each kaizen event should begin four weeks in advance. The length of each event will vary, and you will get better at scheduling events as you gain experience. The key is to set start and end dates and then make every attempt to stay within the confines of that schedule.

Kaizen team leaders must decide when to deviate from the schedule, but for optimum planning, always dedicate resources and energy to meeting the time frames.

Kaizen Team Goals and Expected Results

During the early planning phase of any project, generating goals can be difficult. Forecasting, in any form, can be lucky or lousy. It is important that each kaizen team be faced with some moderate challenges, and these events are being conducted to improve your business, so don't be afraid to set high goals. It is best practice to refer to the established shop floor metrics, as discussed in Chapter 2, as a guide for improvement.

Be sure to establish realistic goals, because setting unattainable goals will serve only to destroy the effort. An attainable goal might be improving productivity 20 percent by reducing waste in a line or a process. But there is no real guide for establishing your team goals. Set goals that you feel are realistic and attainable, and make sure that you plan adequately to ensure success.

Kaizen Event Planning

Traditional kaizen training typically teaches people to perform every single task, from the beginning until the end, during the kaizen event without any planning. Although this practice makes the event action packed, experience has demonstrated that trying to accomplish too much can have a destructive result, often placing kaizen teams in unresolvable situations. To avoid these potential pitfalls, I recommend that you undertake a variety of planning activities four weeks before a kaizen event. During planning, you select the target area as well as the team leader and a tentative list of members. In this way, you can also begin planning specific projects.

Planning involves a variety of items and activities. The team may need to reserve contractors, order supplies and equipment, buy or rent specialized tools and machinery, and conduct waste analysis and time and motion studies. The number and type of planning activities will vary depending on the type of event and the specific goals established for it. Solid planning ensures that the kaizen teams are positioned successfully.

Kaizen Event Budgeting

The final item for consideration when you plan kaizen events is to estimate and allocate the money needed to fund the anticipated improvements. Often, no money is actually spent on an event. However, it is important that a budget be established and funds set aside for the kaizen program. These funds will be used, if necessary, for factory improvements, and should be available to the kaizen team leader.

Kaizen Steering Committee

The **kaizen steering committee** is a governing body of middle and upper managers that oversees all the lean and kaizen initiatives taking place in the factory. This group of decision makers should meet once a month to discuss past, current, and future kaizen events. The group's major charter is to eliminate any obstacles that will impede success, to ensure that the kaizen champion remains focused, and to help decide which events will take place. This committee is also responsible for creating the goals for each team and ensuring that the planning activities are conducted effectively.

Each member of the kaizen steering committee plays a unique role and is required to allocate the appropriate resources for the kaizen program. Although every company can choose its own structure, following is a recommended roster for the kaizen steering committee:

- Kaizen champion
- General or plant manager
- Engineering manager
- Manufacturing manager
- Human resource manager
- Purchasing or materials manager
- Maintenance or facilities manager

Figures 3.1 and 3.2 illustrate the difference between a traditional organizational chart and the kaizen steering committee chart. As you can see, the steering committee has the required managers to assist with decisions regarding kaizen events. Selection of the committee members depends on your needs as an organization. This book focuses on the

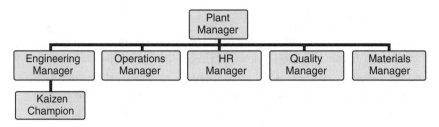

Figure 3.1 Traditional Organizational Chart

factory floor; therefore, the individuals suggested here are appropriate for that focus.

During a monthly meeting, committee members should discuss activities related to kaizen events, schedule events, select team leaders and members, and create goals and objectives for the team. It is a good practice to hold the meeting during the same week each month. Be sure to keep the proceedings organized and the discussion focused on kaizen and kaizen events.

Kaizen Champion

The kaizen champion is the head of the kaizen steering committee. The remaining members are middle or upper managers, but when they meet as a steering committee, the kaizen champion takes the leadership role. Project budgets and schedules are developed by the kaizen champion, and this information is then discussed during the kaizen meeting. The kaizen champion watches or leads all kaizen events, and he is responsible for ensuring that the lean program is being communicated effectively to the factory. The kaizen champion will probably lead many of the initial events while future team leaders are being developed, so the champion knows about all the kaizen activities taking place.

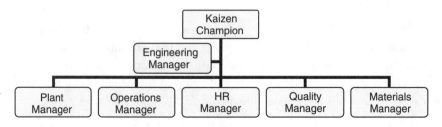

Figure 3.2 Kaizen Steering Committee Chart

General or Plant Manager

Probably the most obvious seat is that of the general manager. Sometimes referred to as the plant manager, this individual is on the committee because it is her plant. The general manager makes most major plant decisions; therefore, she has influence over what takes place within the lean program. I think it is important to allow the kaizen champion to have the most influential power in this committee, because the other members are there to provide guidance. However, the general manager needs to be aware of how each kaizen event will affect the key shop floor metrics and what the possible financial impact may be, because she has final authority to approve the budgets for the events.

Engineering Manager

Engineering must be represented on the kaizen steering committee. The engineering manager has responsibility for most of the manufacturing engineering support staff, including the kaizen champion. The task of the engineering manager is to ensure that the other managers allow the kaizen champion to run the meeting and that focus remains on the lean program. Quality engineers and technicians, industrial engineers, and process engineers typically report to the engineering manager. Therefore, his presence is critical.

Significant changes may be made to line layout, work content, quality checks, testing, and inspection. These activities are considered part of engineering, so the engineering manager needs to ensure that the right people in the department are on the kaizen team.

Manufacturing Manager

The manufacturing manager plays an important role on this committee. Sometimes called the operations manager or production manager, this person must ensure that adequate floor personnel, including operators, are present on each kaizen team. Also, the manufacturing manager must be able to run the day-to-day operations while kaizen events are being conducted. This can be a difficult time, so the manufacturing manager needs to know which improvement efforts are taking place. In addition, the people working under the manufacturing manager, such as supervisors, line leads, and line technicians, will be asked to participate in the events. Any cultural change issues or resistance must be resolved through the manufacturing management department.

Human Resource Manager

When I am conducting my kaizen program training, I get a lot of funny looks when I mention that the human resource manager should be on this committee. Participation on the kaizen teams requires 100 percent dedication, and this is why those chosen will have their usual work given to someone of greater or equal talent. Work can be temporarily given to other people within a department, and the HR department is aware of scheduled vacations, medical leave, short-term disability, and other training that may be occurring. As committee members create the tentative list of team members, the human resource manager can verify that the named personnel do not have any work restrictions or are not scheduled to be away from the factory. Additionally, the kaizen team may be working during off hours, and the HR manager should help make the appropriate safety and security preparations.

Purchasing or Materials Manager

As you make changes to the factory, a variety of items will need to be purchased for implementation. Also, the purchasing manager may have responsibility for the materials department, and her employees can contribute to the kaizen teams with regard to materials presentation, parts quantities, and supplier or receiving issues. During the implementation of 5S and visual management, the kaizen team will need floor tape, labels, tape measures, box cutters, signage, and so on to make their areas 5S compliant. The purchasing department will need to order these items.

As time goes on and your organization becomes more aggressive in its inventory reduction efforts, the purchasing department will become a larger player on the lean journey. New material replenishment systems may be developed, suppliers may change, vendor-supplied material may be implemented, and so on. Purchasing must be heavily involved.

Maintenance or Facilities Manager

A kaizen team leader's most trusted and needed ally is the head of the maintenance or facilities department. Kaizen teams will be relocating workstations, workbenches, computers, equipment, tools, and many other items and machines that need to be fully operational after they are moved. During this transitional stage of a kaizen event, teams must leave the line or area in workable condition for the production run to take place on the next shift, regardless of the location of the line items. All the kaizen committee members should be aware of, and in agreement with, the confusion that may exist as lines are temporarily in flux. Regardless of the event activities, the line operators need to be able to perform their tasks.

The maintenance manager also must provide resources for the kaizen teams for the purpose of connecting wires and pipes, performing network drops, and other activities, as well as ensuring correct and safe functionality. Forklifts will be needed to move large items; therefore, I recommend that all kaizen team leaders become certified to drive forklifts as the kaizen program progresses.

Kaizen Event Tracking and Scheduling

The key success factor in a kaizen program is organization: tracking the progress of the kaizen teams and ensuring that each team meets its established goals. It is also important to follow some guidelines when you schedule kaizen events.

Tracking

The kaizen steering committee needs a visual aid to gauge the success of the program, and it is wise to develop a tracking system for following the kaizen events and helping the teams meet their objectives. A **kaizen event tracking worksheet** is an effective tool for this purpose. It is good practice to allow all employees to view the tracking sheet so that they can follow the progress of the kaizen events and see how they are positively affecting the company. Thus, it's a good idea to implement this tracking worksheet in the form of a spreadsheet that is accessible through the company intranet.

The worksheet is divided into two sections: pre-event items and post-event items.

Pre-event Items

Pre-event items might include the name of the kaizen event, the team leader, the team members, the scheduled dates, the estimated goals and objectives, the key shop floor metrics, planning items, and estimated costs. This information is established during the planning phase for each event.

Post-event Items

When the kaizen event is completed, the kaizen steering committee convenes for its monthly meeting and completes the second portion of the tracking worksheet. These **post-event items** include the results achieved and a comparison to estimated results; action items that have been assigned to the kaizen team members; due dates and the names of those responsible for the action items; and the status of these projects. Kaizen team members should be given no more than 30 days to complete their action items.

Scheduling

To ensure that kaizen events are successful and beneficial to the company, you must make sure that certain tasks are completed; conducting kaizen events haphazardly will not generate positive results. Following are simple guidelines related to timing.

Four Weeks before the Event

- Select the area or line to be improved.
- Appoint a kaizen team leader.
- Create a tentative list of potential team members.
- Order supplies and other items for the team.
- Collect current state data: time studies, waste analysis, spaghetti diagrams, and so on.
- Establish goals and objectives.
- Estimate the cost of the event.
- Identify the team's work area.
- Reserve external resources: contractors, temporary help, employees from sister plants.

Two Weeks before the Event

- Finalize the team members.
- Verify that external resources are scheduled and confirmed.
- Have the team members take a look at the area.
- Analyze the current state for waste-reduction opportunities.

One Week before the Event

- Gather current state shop floor metrics: productivity, floor space, quality, and so on.
- Have the kaizen champion or team leader meet with the team.
- Place kaizen team supplies and necessary items in the team work area.
- Look over the collected data.

One Day before the Event

- The kaizen champion and team leader meet with the plant manager.
- The team leader and team members meet to discuss the team's objectives.
- Verify that all planning activities are complete.

Planning four weeks before the event is a good practice because it avoids overloading the kaizen team with work on day 1 of the event. The team will come up with additional improvement ideas during the event, and that will add to its workload; therefore, scheduling anticipated items in the four-week, two-week, and one-week stages will help the team manage the work effectively and accomplish its goals.

Kaizen Event Communication

Communication regarding the kaizen program and the improvement projects is essential to changing the company culture and keeping the lean program afloat. Comprehensive communication will help employees throughout the company understand the importance of the kaizen program. Posting schedules, team rosters, and past results and accomplishments will help alleviate possible negative attitudes about the program.

Information flow can be difficult if there is no established outlet for delivering it. The kaizen champion is responsible for creating and establishing this important system and keeping the information current and timely throughout the kaizen program. Following are simple, effective suggestions for communication mechanisms.

- Kaizen event tracking sheet
- Kaizen newsletter
- Communication boards
- Employee suggestion box

Kaizen Event Tracking Sheet

Earlier I suggested that this tracking sheet be used to monitor kaizen events. Although this tool is typically used by the kaizen steering committee and the kaizen champion, I recommend that it be accessible to all employees. You can post it via the company intranet or in break rooms. The kaizen champion is responsible for keeping accurate information on the tracking sheet and updating it before every kaizen steering committee meeting. Figure 3.3 shows a typical kaizen event tracking sheet. Notice the categories to be completed during pre-event and post-event activities.

In Figure 3.3, the 520 engine line event is complete. All the rows are filled in, and the action items created after the events are nearing completion. As you can see, there were a few planning items that were done before the start of the event. The purchasing manager was assigned to order new wire racks, possibly for storing material and parts in the new workstations. Time studies were performed to identify waste in the process and improve the efficiency of the line design. This task was assigned to the kaizen champion, as it should be. The facilities manager had to call a local equipment supply company to reserve a scissor lift,

Kaizen Event Tracking Worksheet

Kaizen Event	Date/ Length	Team Leader	Team Members	Planning	KSC Member	Strategic Purpose
520 Engine Line	Week of 8/15/07	Peter Thompson	Jennifer Michaels	Order Wire Racks	Purchasing Mgr.	Productivity Increase
			Albert Stuart	Time Studies	Kaizen Champion	Floor Space Reduction
			Ryan Lee	Reserve Scissor Lift	Facilities Mgr.	Scrap Cost Reduction
			Mike Stone	Verify Vacation Schedules	HR Mgr.	Product Throughput Reduction
			Greg Brown			
			Lisa Albright			
710 Work Cell	9/15/07	Sloan Johnson	Sean Dunlap	Time Studies	Kaizen Champion	Productivity Increase
			Charlie Black	Schedule Floor Painters	Plant Mgr.	Floor Space Reduction
			Erin Kelly			Scrap Cost Reduction
			Colin West			Product Throughput Reduction
			Isaac Relic			

Figure 3.3 Kaizen Event Tracking Sheet

possibly because the factory did not possess one and the team would need it for moving wires and cables connected to the ceiling. Finally, to ensure that all participants were available for the event, the human resource manager was given the task of verifying vacation schedules.

The kaizen steering committee identified the key shop floor metrics that were in need of improvement on the 520 engine line. Each metric in the anticipated results column was assigned a target. An initial budget of $1,000 was developed, to be used by the team during the event for needed items or projects. After the event was completed, post-event information was added, including actual results, event spending, and the action item assignments. The lower portion of the kaizen event tracking sheet in Figure 3.3 is an example of the pre-event information needed to schedule a kaizen event for the 710 work cell.

Kaizen Newsletter

Another effective way to communicate kaizen event information is through a monthly kaizen newsletter. This newsletter is a fun method of keeping employees informed of the improvements made. This news-letter should include the following information:

Anticipated Results	Actual Results	Event Budget	Event Spending	Action Items	Responsible	Due Date	Status
20%	28%	$1,000	$975	Update Work Instructions	Albert Stuart	9/10/07	Complete
50%	45%			Install Tower Lights	Greg Brown	9/18/07	1/2 Complete
90%	95%			Install Tower Lights	Greg Brown	9/18/07	1/2 Complete
40%	50%						
20%		$2,300					
25%							
70%							
35%							

- Upcoming kaizen events schedule
- Kaizen team members, with photos and individual accomplishments
- Goals and objectives
- Results of previous events, with "before" and "after" pictures

The newsletter can be distributed with employee paychecks, placed in break rooms and at the front entrance, or mailed to the employees' homes. Be creative in the development of the newsletter, and encourage team members and leaders to write articles. The newsletter is an effective and fun approach for promoting culture change and encouraging everyone's involvement in the kaizen program.

Communication Boards

When placed in the right location, dry erase boards or chalkboards are equally effective in displaying kaizen event information. Break rooms are ideal locations for the boards, because the operators and other floor personnel will see them whenever they take breaks and eat lunch. Employee entrances, including back or side doors to the factory floor, also are good locations. Communication boards can also be placed in meeting and conference rooms so that office employees, engineers, and managers can view them. Most of the information published in the newsletter also should be posted on the communication boards. Make certain that these boards are used only for kaizen event information.

Employee Suggestion Box

Line operators are often left out of the design and planning phase of kaizen events. I have discussed the importance of including operators on kaizen teams, and they should be involved in deciding which area is scheduled for a kaizen event.

Operators are usually confined to their workstations or areas on the production floor and typically get little or no contact with management or engineers. Any contact that occurs is usually initiated by the support staff in the work area. So how can you get operators and other floor employees involved in the decision-making process and get their input for continuous improvement? I recommend the use of an employee suggestion box, which allows operators to give input on future improvements.

Kaizen Event Suggestion Form

Name_____   Date_____

Line or Work Area Assigned_____

Kaizen Event Idea_____

Area Suggested_____

Would You Participate? Yes_____ No_____

Thank you for your suggestion. The kaizen steering committee will
review your request at the next kaizen meeting.

Figure 3.4 Kaizen Event Suggestion Form

Much like a ballot box, the employee suggestion box is used to collect
ideas from line workers about future kaizen events. The box should be
placed near the communication boards or in operator break rooms. A
simple suggestion form like the one in Figure 3.4 should be placed near
the box.

The kaizen champion should remove the suggestions every week, pre-
senting them to management at the next scheduled kaizen meeting.
Every suggestion should be reviewed and considered for future kaizen
events. Keep in mind that some employees will use the event suggestion system to complain about issues outside the kaizen program. Make
it clear to employees that the box is for kaizen event ideas only. They
should also have the option not to reveal their names, but remember
that you are trying to get them involved beyond the suggestion phase. If
the idea is selected, you want the individual who conceived the idea to
implement it as well.

Monthly Kaizen Meeting

Each kaizen meeting should last no more than an hour and should
include the following three agenda items:

- Open action items
- Past kaizen event results and lessons learned
- Upcoming kaizen events

Open Action Items

To ensure that kaizen event action items do not go unfinished, team members with incomplete action items must attend the first part of the meeting. With the kaizen event tracking sheet displayed, the kaizen steering committee and the team members can discuss the progress of their projects. The committee's role is to clear any obstacles or resolve issues that hinder completion of an item. This part of the meeting is not an opportunity for finger-pointing but rather a chance to see whether there are legitimate reasons the action items are not complete. Of course, the team members can simply provide the status of their tasks if the deadline for completion has not yet arrived.

Past Kaizen Event Results and Lessons Learned

The second part of the kaizen meeting is to review the line's progress. The kaizen team leader for each event being discussed should be invited to participate. Team members who were invited only for part 1 can now return to work.

In this portion of the meeting, the following questions are addressed: Are the operators up to speed? What kind of output are they producing? How is productivity? Was the estimated floor space reduction truly accomplished? Now is the time to be candid and to learn everything about the prior event. Each kaizen event will present a different learning curve after implementation, when operators, supervisors, and support staff must adjust to new standards, procedures, and, more than likely, a physical process that is different from what they have been used to.

Lessons learned also can be discussed. The team leader should be given the opportunity to provide input on the positive and negative aspects of the event and discuss them with the committee. These aspects could include team participation, the accuracy of the data collected, the specific time of day that the event was held, the kaizen steering committee's planning activities, and so on. It is essential for everyone to be honest so that you can conduct better and more successful events in the future.

Upcoming Kaizen Events

As the third part of the meeting begins, the previous team leaders can return to work and the kaizen committee can begin to plan future events. Remember to follow the kaizen event planning rules with regard to scheduling the event, selecting the team leader and members, and assigning the required planning projects to the committee members.

Your company's kaizen program is essentially a continuous improvement infrastructure that acts as a catalyst for your lean program. Each element is critical to the program's long-term success; therefore, all new employees should be made aware of the program immediately. New managers can be informed of the program during the interview process and can be made aware that they could be selected to be on the kaizen steering committee.

There will be some initial resistance to this program, because it may be perceived as another "flavor of the month." In the beginning, it may be difficult to schedule and conduct monthly kaizen events. Many companies have other important projects running in parallel, so it may be best to schedule kaizen events every other month rather than monthly. Still, companies for which I have been a consultant and trainer have reached a point where they can successfully sustain two or three kaizen events each month. Of course, sustaining multiple events simultaneously requires multiple kaizen teams and many kaizen team leaders. Running more than one event at a time is a challenge, but, with time, as the kaizen program is strengthened and processes become streamlined, chaos is significantly reduced.

Getting Started

There is always a lot of excitement at the start of a kaizen program, and many people want to dive in immediately and get started. To avoid mishaps and ensure that you don't try to tackle too many activities at one time, I provide this list of a few activities you may wish to consider in the early stages of the program:

- Schedule a meeting with upper and middle management.
- Schedule and conduct the first kaizen meeting.
- Create the kaizen event tracking sheet.

Schedule a Meeting with Upper and Middle Management

The most effective first step is to bring together the key managers who potentially constitute the kaizen steering committee. Depending on your organization, invite the plant manager, engineering manager, operations manager, purchasing manager, human resource manager, and facilities manager. Discuss the development of a kaizen program, and outline its importance to the company's lean manufacturing initiatives. Explain that a structured company kaizen program, with ongoing kaizen events, is vital to a successful lean journey. Outline the kaizen steering committee, and describe how each manager in the committee is responsible for allocating the necessary resources (time and personnel) to each kaizen event. Attempt to finalize the committee membership and schedule the first kaizen event.

Schedule and Conduct the First Kaizen Meeting

Because the kaizen champion has not yet been appointed, the plant manager or engineering manager should lead the first meeting. Selection of the kaizen champion is the first order of business. Begin by describing the role of the kaizen champion and explaining that this individual will be dedicated 100 percent to lean manufacturing initiatives and will not be required to perform other work. Indicate that the kaizen champion should be a solid performer and that this work should not be taken lightly.

Depending on current available resources, is there someone in the company who can perform this role? The kaizen committee needs to review the candidate's current job and decide how her current responsibilities will be delegated. If you will hire someone from outside the company, discuss the job description.

Next on the agenda for the newly created committee is the discussion of future kaizen events. The first kaizen event does not necessarily need to be scheduled, but the committee needs to review the program rules with regard to scheduling, planning, and team selection. Assign the various kaizen projects to the committee members, such as building the suggestion box, developing the communication system, and creating the kaizen event tracking sheet. Once these items have been assigned, schedule the second meeting, which will be used to begin planning and scheduling kaizen events.

Create the Kaizen Event Tracking Sheet

The kaizen event tracking sheet is usually designed and maintained by the person who will eventually be the kaizen champion. I have provided a sample kaizen event tracking sheet that I have used many times (shown earlier in Figure 3.3). If necessary, modify it to your own program. This tracking sheet will be used for all kaizen event activities, so I suggest that it be ready before your second kaizen meeting.

Your First Kaizen Event

During the planning weeks leading up to the first kaizen event, all the fundamental elements of the kaizen program should be put into place. The key to your first event is simplicity. Do not embark on a high-profile assembly line redesign that includes previously conducted time and motion studies, massive waste reduction, 5S, and new line layout and workstation calculations; this kind of work would be far too comprehensive to accomplish successfully at the beginning. This does not imply that the talent is not there to make it happen but simply that being too aggressive in the first event can be dangerous.

A 5S kaizen event is probably the smartest event to begin with. Start small, and schedule an event that will be used to implement the 5S philosophies in one area. 5S is very common and is applied even in companies that do not have a lean manufacturing program in place. 5S is a methodical approach to cleanliness and organization, allowing you to identify and remove unnecessary items from an area, clean the area, put things in place with proper designations and identifications, and then maintain the process for consistent compliance.

Select a small area, and schedule the 5S event to last approximately two or three days. Develop reasonable goals for the team, and then communicate the first kaizen event with excitement. Your goal in the first event is to get a quick win and to motivate employees to do more. Following is a typical three-day 5S kaizen event that illustrates how to plan and execute this activity.

Planning

5S events do not require as much up-front planning and data collection as, for example, preparing for a new line layout. The purchasing manager on the kaizen steering committee should order supplies:

- Floor tape: to be placed around items on the factory floor
- Labels: to identify parts, tools, and other necessities in bins
- Laminating machine and sheets: to protect signs and designations

- Paper cutter: for use in cutting neat, square signs and designations
- Plastic bins and parts racks: to store materials and parts in workstations

This list can be much longer. This one is designed as a starting point for your first 5S kaizen event.

Selecting the Area

You should choose an area that is small and can provide good results, creating enthusiasm for future kaizen events. You might choose a sub-assembly work area that may consist of a few workstations; a small main assembly line or a work cell with about five workstations; or a packaging area. It is also best to select an assembly process or other manufacturing area that is building product, rather than locations such as receiving or the maintenance department. By selecting an area on the production floor, you ensure that many people will be exposed to the first kaizen event.

Selecting the Team Leader

If you have chosen your new kaizen champion, use this person as team leader. Otherwise, appoint a senior-level manufacturing or industrial engineer to lead the first kaizen event. Make sure that his normal day-to-day responsibilities have been reallocated for the duration of the event. It will be a learning experience for this first-time leader, so make sure he has everything needed to be successful.

Selecting Team Members

Because this first event is to be conducted in a small area, select five team members:

- Line engineer: Understands the process and products.
- Production supervisor: Understands the current culture and will be the enforcer of the new lean tools.
- Member of the kaizen steering committee: Always good to have a manager on the first kaizen team.
- Line operators: Have in-depth knowledge of the product and how it is built. What's more, the implementation must include them to ensure buy in and to help teach and train the operators on the line after the event is complete.

Week of the Kaizen Event

A variety of tasks takes place each day during a kaizen event. Here is a recommended guideline.

Day 1

The first day of a typical 5S kaizen event is dedicated to removing all unnecessary items from the work area. First, take pictures of the line or area to be used for comparing its current state to its future state.

Second, divide the team into two subteams: a sort team and a receiving team. The **sort team** is in charge of identifying the items that are no longer necessary for the operators—for example, workbenches, tools, parts, documentation, chairs, and garbage cans. Each item should be properly identified with a tag, often referred to as a **red tag.** A red tag is filled out and placed on the object to be removed.

The line operators on the team should be part of the sort team because they have firsthand knowledge of what is needed and unneeded. Ask a lot of questions as a group to ensure that the right things are being tagged. Leave only the items needed to do the work.

The **receiving team** should set up an area in the factory for accepting tagged items from the sort team. Place signage identifying the area as off-limits to the other employees. As the items begin to arrive in the collection area, the receiving team should document everything on a master list, keeping track of the discarded items. Toward the end of the day, the sort team should wrap up its objectives and can assist the receiving team in documenting the final results. It is good practice to invite the kaizen steering committee to the collection area to view the mountain of unused tools, parts, workbenches, and so on that was identified as unneeded and removed from the area.

Days 2 and 3

The sorting on day 1 has created an island effect, so days 2 and 3 are dedicated to putting the line or area back together. You will be surprised by what has been identified as truly needed to perform the work. As the team starts to compress the line and put the remaining items in order, start creating designations on the floor.

Using floor tape, identify each item—such as pallets of material, parts racks, garbage cans, and so on—with its appropriate floor designation.

Each item should also have a label on the floor identifying what it is and possible quantity requirements. Put parts into bins, and label each bin with the correct part description, part number, quantity, and workstation.

Create workstation signs and other designations to provide a visual guide to the workstation design. Allow the team to be creative and come up with interesting ways to identify items. Everything should have a home, including miscellaneous tools and supplies. Make sure that tools and air hoses are not on the floor and are placed at waist height or overhead. Finally, clean the area, giving it a showroom appearance. The team leader should take pictures of the final layout showing the improvements made by the team.

Last Day

The team leader should put together a presentation that will be shown to the kaizen steering committee and other employees either at the end of day 3 or the following morning. It should contain before and after pictures, along with the individual accomplishments of each team member. Finally, take a tour of the line or area, including the collection area.

Chapter Wrap-Up

Developing the individual elements of the company kaizen program is not difficult, but it takes time. Remember the following key points: (1) Every successful lean program requires a firm foundation that allows a company to allocate the appropriate resources and make time to implement the lean initiative; (2) you should establish an effective communication system that will ensure buy in, participation, and awareness; (3) you should allow every employee to have an opportunity to participate in a kaizen event and to offer input and suggestions; (4) the kaizen steering committee, dedicated to planning and execution of kaizen events, is an essential component. Your lean journey will progress with this company kaizen program.

four

Early Stumbling Blocks

So you have sent your engineers and middle managers to the best lean manufacturing training money can buy. Now what? It is a lot of information to absorb, and there is a tremendous amount of work that must be done. Chapter 3 explained the importance of having a comprehensive company kaizen program to help you plan and execute kaizen events to implement lean manufacturing. In this chapter, I explain the fundamental building blocks of lean manufacturing as they apply to 5S, data collection, quality, and workstation design.

Attention to detail ensures the success of a lean journey. Not all the tools and techniques in the lean philosophy are applicable to all processes, so it is important to learn about the tools and apply them where they are appropriate.

In this chapter I also outline the common mistakes organizations make when preparing information for new process designs. It takes time to learn and use most of the tools and techniques effectively. Therefore, explaining some of the stumbling blocks in the early stages of the journey should help you avoid failure of your lean implementation. Here are the basic tools:

- 5S and the visual workplace

- Time and motion studies

- Waste removal
- Quality at the source
- Workstation design

5S and the Visual Workplace

The 5S organization and cleanliness philosophy is the cornerstone of any manufacturing environment. The 5S system is a powerful continuous improvement tool that can generate immediate results. Here again are the five principles:

- Sort: The act of removing and discarding all unnecessary items from the work area.
- Straighten: The act of organizing what is needed so that it is easily identifiable in a designated place
- Scrub: Cleaning the area so that it is in showroom condition
- Standardize: The act of following best practices and maintaining consistency in the work area
- Sustain: Maintaining the organization and continually improving on it

I have been involved in many 5S implementations. It takes some time to be implemented factorywide, but 5S is a simple program that is easily adopted. It can dramatically improve productivity, floor space use, throughput time, and cost.

The most common mistakes occur in straightening and sustaining, and therefore I will focus on these two areas.

Mistakes in Straightening

Your production floor is your showroom. When suppliers, customers, and investors tour the floor, the state of the work environment—clean or disordered—is a reflection on the company's work ethic. Tools on the floor, disorganized paperwork, tables and workbenches out of place, and dirt and grease on the floor all present a bad impression; it does not look professional. Workplace pride is critical to success, but it is difficult to cultivate an atmosphere of pride when the environment is cluttered, dirty, and disorganized.

The sort function of 5S is easy to understand and implement. In my experience, most operators, engineers, and managers can easily discard unneeded items and even feel good about it. Clearing clutter can be invigorating.

The first mistake depends on how the **straighten** task is interpreted. The idea of 5S is to bring out your employees' creativity and talent, allowing them to create unique and effective means of organizing the workplace. People do not always do as much as they can in this area, and that is a mistake. You need to identify everything on the floor: workbenches, tools, parts, parts bins, station signs, storage areas, documentation, garbage cans, chairs, and so on. If the item is necessary in the work area, identify its correct location and then designate it with plenty of labels, signs, insignia, and colors. Hold nothing back.

Imagine your factory having a place for everything. Imagine every item clearly marked with its name and its quantity. Figure 4.1 displays the level of detailed identification required for a garbage can that is located in a workstation on the factory floor.

It isn't sufficient merely to place the garbage can somewhere on the factory floor. First, you should identify its location using a particular color of floor tape (yellow is the most popular). Next, label the garbage can. Designating it as a "garbage can" is not enough, because there are numerous garbage cans in the factory. In the example, the garbage can in Figure 4.1 is from the G line and is marked accordingly.

A story from my experience illustrates why this level of detail is important. I was helping a company in Nebraska with a series of kaizen

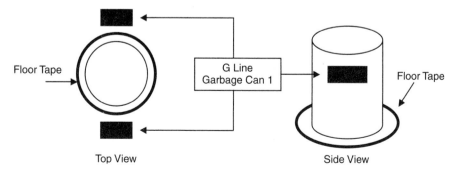

Figure 4.1 Garbage Can Identification

events. We began by implementing 5S on one assembly line. Three line operators participated on the kaizen team. Day 1 of the kaizen event was designated for sorting all unnecessary items on the line. It was a small line, so the team completed the sorting portion of the project in a couple of hours. By the end of the day, the team was putting together the new line.

One operator was eager to help and was working with me in a particular workstation. I allowed him to come up with ideas regarding placement of the items in the station. The workstation required a garbage can, because some of the parts arriving from suppliers were protected with foam that had to be removed and discarded. The team tried to find a way to reduce the amount of unpacking that was done on the line, but this particular part was painted and needed to be protected until installation.

The operator decided that the garbage can should be located to the left of the workstation. He had participated in the 5S training that I had conducted, so he knew I wanted him to be creative. The garbage can was placed on the floor at the end of the day, and he decided to finish his 5S plan the following day. When we returned to the workstation the following morning, the garbage can was missing. I was not surprised that it was gone. He found it comical, because there were hundreds of garbage cans in this factory and they were always disappearing. Laughing, he ran off to find his garbage can.

Upon his return, I asked him why he thought it had been missing. Fortunately, his answer was "poor identification." The operator placed the garbage can back where it was supposed to go and then proceeded to place yellow floor tape around it. Because there was a lot of work to do on the line, we left to help other team members with their implementations.

I did not see much of him that day except in passing. I looked for his garbage can, and it was still in its place. On the third day we returned to the location and, lo and behold, the can was gone—just as I knew it would be. Essentially, I was allowing the operator to make some key mistakes, helping him grasp the concept of 5S through his own experience. The operator now seemed a bit disgruntled, but again he was eager to solve the issue. He returned the garbage can to its place, placing it in the middle of the floor tape. He then placed a label on the floor that read "garbage can."

I knew that was not enough, but I wanted him to learn a little more. By the end of the day, the can again had disappeared, this time getting only as far as the next workstation. Because the event was nearing completion, I felt compelled to intervene. I explained that because there were numerous garbage cans in the factory he might want to change the label on the floor to "G Line Garbage Can 1." This would ensure that the garbage can would stay in the G line and in workstation 1. The garbage cans in the other workstations on the G line would be labeled 2 and 3 to ensure that they remained in their own stations. He got it! I apologized for dragging him along, but I explained my reason for wanting him to learn from his own mistakes and experience kaizen in a way that would make him a true believer.

The same numbering approach can be used for brooms, dustpans, mops, or any other item that has multiple quantities for a given area. Make sure that the item is labeled with its name in addition to its location.

My example is just that—an example. It's a good idea to tap in to the talent of your employees, allowing them to devise the best, most creative manner of organization. Here are a few more examples.

Figure 4.2 shows a simple, effective way to identify a workbench. In this example, the workbench is designated as T1. ("T1" distinguishes this workbench from other workbenches. A designation of "workbench" is not sufficient because there may be more than one workbench. The best analogy is a house number and street name. Imagine trying to get mail without an address or if the house were simply called "house.") In Figure 4.2, floor tape has been placed around the workbench, and the workstation sign is posted above it for easy viewing. The same workstation identification is on the floor next to the floor tape. From most vantage points, this workstation is highly visible.

Workbenches, stations, shelves, and parts racks need identification because manufacturing items such as tools, parts, materials, supplies, and documentation also need a home location. These items are generally placed in a bin or on a flat surface in a workstation. T1 would have a collection of necessary items for performing the work, and each item's location in the factory is in T1.

The bins or containers holding these parts should also be clearly marked and designated as belonging in T1. The label should include the part description, part number, part quantity, and location (see Figure 4.3).

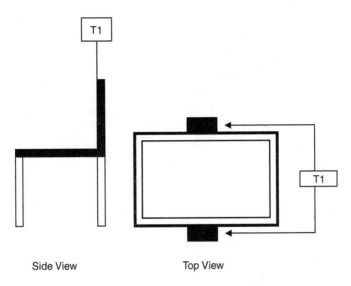

Side View   Top View

Figure 4.2 Workbench Identification

The contents of parts bins should be clearly marked with their correct location. This ensures that a "wandering" part returns to its home location on the production floor. In this example, the location is T1. A similar label should also be placed on the workbench T1. In this way, the parts bin sits in a designated place on T1.

Try to be as creative and detailed as possible when you are straightening.

Mistakes in Sustaining

In most cases, organizations successfully implement the first four S's in the 5S program. Sustaining the changes is a different story. Culture change plays a major role in its success. Holding people accountable can be difficult, especially when the employees have been with the company for quite a while or have been performing the same job function for many years. It is human nature to grow comfortable with

> Alternator Bracket
> 90-76554-22
> Qty: 15
> Location: T1

Figure 4.3 Part Identification

established ways of working and to want to maintain that comfort zone even after physical changes are made to the work environment.

Employees who are resistant to change will battle against the new requirements for order and cleanliness as long as possible. To combat this, from a managerial perspective, you need to have in place two monitoring systems:

- 5S audits
- 5S tracking sheet

5S Audits

Auditing and tracking any procedure or program are critical to its success. You cannot simply put a new program in place and expect it to work on its own. This is especially true with lean manufacturing. To ensure its sustained success, you must maintain it and continuously improve upon it. Your 5S program will be successful if it has the commitment of those on the floor and if management audits and maintains the area consistently.

You must create a **5S audit form** that captures all the important elements of organization and identification for the specified area. For example, the criteria you establish for assembly lines will vary from the criteria for the fabrication or receiving departments. Each area will need its own 5S audit form. Also, you cannot perform an audit until after the company has implemented 5S in that area. Although this seems obvious, some people attempt to do it backwards.

Figure 4.4 shows an example of a 5S audit form for production or assembly lines. You need to tailor your forms to your processes and to your goals in monitoring your 5S program.

The 5S audit form is used to monitor conformance to the company's 5S program. The form is based on the five S's (sort, straighten, scrub, standardize, and sustain). You need to establish and then monitor the various criteria. There is no maximum number of criteria; you can list as many as you'd like.

The 5S form can be short or lengthy. I have seen forms that were four pages long. The scoring portion of the audit can be established in two ways: a yes or no approach, or scoring based on a scale. The form in Figure 4.4 uses yes or no answers to indicate conformance.

5S Audit Form

Team				
Audit Date	# of Yesses	/15 =	%	
Auditors				

Sort (Get rid of unnecessary items)

Workstation and/or area is clear of all non-production required material.	Yes	No
Unnecessary equipment and machines have been removed from the area.	Yes	No

Straighten (Organize)

Items on the floor are clearly marked with floor tape and labels.	Yes	No
All equipment and tools are clearly marked and well organized.	Yes	No
Locations and containers for items, parts, and supplies are clearly marked.	Yes	No

Scrub (Clean and solve)

Floors, work surfaces, equipment, and storage areas are clean.	Yes	No
Garbage and recyclables are collected and disposed of properly.	Yes	No
Excess pallet and packaging materials are cleared out of area.	Yes	No
Machines and equipment are free of grease and dust.	Yes	No

Standardize (Tasks)

Standard work is displayed.	Yes	No
It is obvious through visual management whether tasks have been done.	Yes	No

Sustain (Keep it up)

Standard work is being followed.	Yes	No
Work instructions are displayed with correct revision.	Yes	No
Work area is clean, neat, and orderly with no serious unsafe conditions observed.	Yes	No
End-of-day cleanup procedures are posted.	Yes	No

GREEN = 81% to 100% **YELLOW** = 66% to 80% **RED** = 0% to 65%
Area is 5S compliant Area meets minimal standards Area needs immediate attention

Figure 4.4 5S Audit Form

Alternatively, you might want to use a scale of 1 to 5 (5 = complete conformance, 3 = minimal conformance, and 1 = nonconformance).

The 5S audit form should be developed for every process and department in the factory and should be process-specific. As mentioned, the form for assembly lines should be different from the forms for receiving, maintenance, and shipping. These areas operate within different process parameters, and therefore the 5S audit must be tailored to the needs in each area.

For consistency, each 5S audit form should contain the same number of questions. Although each team has different work processes, it would not be fair to audit the maintenance department with 8 questions and then audit the manufacturing lines with 25 questions. It is typically best practice to form different 5S teams for the different areas, because workers from each area know it best and can help establish the criteria and perform the audits. Make sure that the 5S audit forms are developed based on what is considered critical to organization and what is considered the standard for cleanliness in each respective area.

Many companies struggle with developing their approach to 5S and specifying how it should be implemented in each area. It is wise to begin by developing the 5S audit form for the area. When that is completed, each kaizen team will know the specific criteria to use for the implementation, and that almost guarantees that the area will be 5S compliant. Remember, 5S compliance is defined differently for each organization. Develop the criteria, implement 5S, and sustain it.

To maintain consistency, 5S audits should be conducted once a week. It is a good idea to post the audit criteria in the target area so that operators and floor supervisors are visually aware of the standard. After the audit, the data from the audit form is added together, and the completed audit is based on a percentage, as shown at the bottom of Figure 4.4.

There are no secrets here; you want the floor personnel to know what they will be audited on. The company should want the areas to score well on a consistent basis.

5S Tracking Sheet

Four audits should be performed every month (one per week), and the results posted on a tracking sheet. This **5S tracking sheet** needs to be visible to the whole organization, because office 5S audits will eventually become part of the 5S process. Each department, line, work area, or function will be aware of the monthly progress. This is not a finger-pointing game but rather a visual indicator of trends, something that helps management pinpoint areas in need of improvement.

Each month, the department or area with the highest 5S score is awarded a prize. This process creates healthy and fun competition for everyone, and it encourages everyone to strive for the highest score. The 5S tracking sheet should be updated only once a month, and then again at year end to determine the overall winner for the year. However, before you can begin tracking 5S scores, 5S must be implemented plantwide based on the criteria established in the audit form.

Time and Motion Studies

All lean decisions are based on data, and good reliable data will never fail you. It takes time to collect good data, and I am not an advocate of rushing the process. For your lean efforts to positively affect key shop floor metrics—and ultimately to improve cost, quality, and delivery—

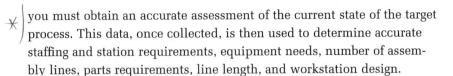

you must obtain an accurate assessment of the current state of the target process. This data, once collected, is then used to determine accurate staffing and station requirements, equipment needs, number of assembly lines, parts requirements, line length, and workstation design.

The most fundamental aspect of data collection is conducting **time and motion studies** for the target process: evaluating the work content, work steps, and movement of line workers for the purpose of identifying value-added and non-value-added work. Collecting this information takes time. Here are some suggestions that will reduce the possibility of errors during collection:

- Use a stopwatch.

- Document the work from start to finish.

- Collect the work content first.

Use a Stopwatch

Historically, time and motion studies have been performed in a variety of ways or not at all. Unfortunately, either many manufacturers do not have this information in their possession, or the data they possess is inaccurate or outdated. Some manufacturers have performed the studies and have accurate data but have never used it. In any case, the data is important and I highly recommend that it be collected and put to good use. My professional preferences for data collection are time and motion studies, because they provide highly detailed information that is invaluable for improving cost, quality, and delivery.

Video cameras can be placed on the floor, focused in the general direction of the work area or inside the workstation. Although these cameras are perfect for capturing excess movement (walking) and excessive wait times, it is often difficult to catch each individual step the workers take while performing their job process. The clock on the camera runs continuously, and therefore the data collector reviews the recording and manually establishes and records the individual work times. A simple wristwatch can also be used to collect times associated with each work step, but, in my experience, it does not reflect actual time.

The best tool for time tracking is a stopwatch. By setting the watch to either seconds or minutes, you can accurately capture the true time it takes to perform a specific part of the process. At least six to eight samples should be collected for each task so that a solid average time can be calculated. This average takes into account any minor differences that occur in each sample.

Document the Work from Start to Finish

The data collector should document the individual work steps from start to finish, beginning with the first workstation and proceeding down the line or process. The goal is to create a long, sequential list of work in progress, capturing every step that is taken. Figure 4.5 is an example of what this data may look like.

To create an accurate picture of the current state of the work on the line, the observer must document every action. This action reveals the processes and their current inefficiencies, such as operators searching for tools and parts while performing the work. The information will indicate actions such as how often the operators leave their workstations and whether they must share tools while performing the work, or walk to a central storage area to find parts and documentation. When all steps are accurately captured, they must be identified as either value-added or non-value-added. Therefore, it's critical to capture every single action.

Work Description: Main Assembly Line				Time Samples					
Step	Work Content	VA	NVA	1	2	3	4	5	6
1	Place Unit into Workstation		X						
2	Install Wire Harness	X							
3	Place Wire Harness Cover Over Wire Harness	X							
4	Secure Wire Harness Cover with (2) Nuts	X							
5	Connect Wires to Terminals	X							
6	Walk and Retrieve Right Hand Panel		X						
7	Return to Workstation		X						
8	Install Right Hand Panel	X							
9	Walk and Retrieve Left Hand Panel		X						
10	Return to Workstation		X						
11	Install Left Hand Panel	X							
12	Place Warning Label on Right Hand Panel	X							
13	Secure Wires with Tie Wraps	X							
14	Walk and Retrieve Wire Tie Cutters		X						
15	Return to Workstation		X						
16	Trim Wire Ties	X							
17	Return Wire Tie Cutters		X						
18	Return to Workstation		X						
19	Move Unit into Queue		X						

Figure 4.5 Work Content List, Main Assembly Line

Collect the Work Content First

Data collectors make mistakes when performing time and motion studies if they attempt to track too much at one time. It is nearly impossible to identify and document work steps and simultaneously time the events, especially for tasks that take less than eight seconds. The only good information is accurate information. Therefore, you must take sufficient time to perform these studies.

At the beginning of the exercise, leave the stopwatch in the desk and concentrate on recording the actual work steps without worrying about the time. During the day, production workers perform numerous activities, and you need to observe all of them. There may be line stoppages or changes in the product mix (on mixed-model lines), or the operators may change the order in which they perform tasks. Before beginning your task, approach the workstation and speak with the line operators. Explain that you will document their work steps in an attempt to learn the process on the line. Because you will not have a stopwatch, they are less likely to be nervous and will probably work at a productive pace. Ask them to maintain consistency in their steps so that you can collect accurate data.

It is important for the data collector to develop a positive relationship with the line workers, for a few key reasons. First, the data collector will be returning with a stopwatch, and therefore a good relationship established beforehand will serve to ease operator stress about being timed. Second, the collected information will be used to improve the process, and therefore it must be accurate. Starting off on the wrong foot will only create a sense of animosity, and line operators may be less inclined to be helpful. Third, creating a sense of teamwork and camaraderie with the operators is important because they will be participating in the kaizen events. Additionally, a line operator can be of great assistance with the initial data collection, helping the data collector gather all important area information before the time studies.

Once the work content has been documented, the data collector can begin the timing portion of the study, starting at any point in the build sequence. The data collector can time any area in any order. For example, she can start at the beginning of the process and then halt the study, if necessary, and move to another workstation. Flexibility is important, because, in reality, an issue on the line could force an operator to stop working for a brief period. This is another reason it is important to document the work steps first before attempting to capture time. By using these guidelines, a data collector can gather all the information efficiently and accurately.

Waste Removal

Waste removal is the cornerstone of any lean journey. It is a continuous process of analyzing and providing solutions for each individual process in the company. In short, it never ends. I have often been asked, "When are you waste free?" My answer is, "Never." Time and motion studies are an essential waste analysis tool when it comes to assembly lines, work cells, machine work, and setup reduction. Although these studies are not the only tool available, they are one of the best. I have just described how they should be conducted. Now, what happens after time and motion studies have been done and the data collected?

Waste is evaluated based on its impact on the process. Improvement teams often make mistakes in their approach to waste removal. Removing some kinds of waste is difficult or nearly impossible, and perhaps should not be attempted. Instead, you should direct your efforts toward high-impact waste according to its priority and its effect on the process overall.

On a flight from Chicago, I read an article about a company that was 100 percent waste free. The article described the organization's lean journey, which incorporated "green" manufacturing into its operations. All manufacturing and office processes, including the construction of the facility, were designed with the environmental footprint in mind. It was a very progressive-thinking company. However, I struggled with the concept of 100 percent waste free. Even though I am knowledgeable on green manufacturing, I found it difficult to grasp the concept of 100 percent waste free, especially with regard to inefficiencies.

In all my years in the lean field, I have yet to find a company operating in a 100 percent lean state. In my opinion that state does not exist. Lean is a journey that never ends. Waste will always exist, and companies must remain consistent and diligent in their continuous attempts to reduce or eradicate it.

With that said, your waste-reduction efforts should be centered on the big hitters. There are essentially three priority levels of waste: low, medium, and high. Kaizen teams should focus on the high-priority waste and work their way down to medium- and low-priority waste.

High-Priority Waste

High-priority waste should be identified and removed from the process as quickly as possible. I classify the action of an operator leaving a workstation as a high-priority waste in both action and associated time. By definition, operators are considered value-added because they build the product that brings profitability to a company. Their time must be focused and directed toward the work taking place in their stations. Walking around and searching for tools, parts, supervisors, documentation, standards, supplies, or any other item in need is high-priority waste. It has an immediate negative impact on the shop floor, creating bottlenecks as well as animosity between production workers. In addition, operators who frequently leave their work areas may lose focus and can possibly commit quality errors.

In my travels, I have been confronted by managers who did not want to add the cost associated with having materials handlers bring parts to workstations. Their argument may initially appear to be valid, because labor costs increase whenever workers are added. But placing materials and parts at the point of use (i.e., in the workstation) definitely results in a reduction of waste and a value-added activity, and this practice ultimately has a positive impact on cost, quality, and delivery.

At this point in the debate, my statements may seem to resolve the dilemma. Not yet. Often, the next management solution is to place an entire pallet or large bin of material inside the workstation to satisfy the need for parts replenishment. Unfortunately, this approach adds more to waste and cost than simply adding a materials handler to the process. Using this method, the organization incurs higher inventory costs and physically increases the size and length of the process. Storing large amounts of material requires more floor and workstation space, adding to delivery time, increasing the cost of doing business, and potentially even affecting quality. The solution to parts replenishment on the line is not to place large amounts of material within the manufacturing process. Nor is it to allow operators to gather their own parts. The more sensible resolution is to provide materials to the workstations in reasonable quantities and to employ materials handlers to deliver the goods as needed.

Not convinced? Consider the example of a dentist. When a dentist works on your teeth, he has an assistant who hands him tools and attends to your physical comfort. This practice allows the dentist to focus solely on your teeth and the important work he is performing. All the necessary tools and supplies are placed on a nearby tray within easy reach of the dentist and the assistant. The dentist and the assistant have no need to leave the patient to retrieve tools or other necessary items. In fact, leaving the patient during an important procedure would not be optimal performance or adequate patient care. Similarly, when line operators leave their workstations, they are practicing high-priority waste. Improvement teams should focus on eliminating high-priority waste as their first action of business. Leaving the workstation should be eliminated, and it usually can be accomplished with point-of-use parts, tools, documentation, and an adequate signal system to communicate operator needs like a tower light.

Medium-Priority Waste

Medium-priority waste is generally associated with waste that occurs while operators are within their workstations. Operators are often required to remove parts from original supplier packaging. Removal of foam, shrink-wrap, tape, bags, and boxes can consume valuable assembly time and requires garbage cans or recycling bins in the production area. In regard to the seven deadly wastes, this would be considered overprocessing. Packaging should be removed in the receiving area unless the parts need to be protected from damage. It is a case-by-case situation, of course, but most parts can be delivered to the stations ready to install.

Two other examples of medium-priority waste are waste caused by imbalances in the work flow or from poor communication. Assembly lines should be well balanced to ensure a steady flow throughout the workday. During wait times, operators should be working together and flexing within the workstations, but this practice still does not solve the waste issue. To reduce or eliminate any problems in work flow, you should perform accurate time studies to evaluate the optimum work content balance.

Lack of timely information or insufficient communication can also cause unnecessary wait times. Operators trying to locate a production supervisor must have appropriate methods and signals to communicate their need. Without adequate ways to communicate, they may opt to leave their workstations, placing them in the high-priority waste category. Operators who cannot communicate their needs may continue to work without answers to their questions or without necessary supplies until the appropriate person happens to stop by the workstation. In some cases, they may need to stop work completely and just wait until someone comes by to address their need.

Poor workstation design can create wasted motion, which I also define as medium-priority waste. Lack of 5S and organization can make it difficult to locate items in the workstation. Workers can become frustrated if labels are difficult to read, part numbers are hard to distinguish, or torque values cannot be identified on tools. Poor tool installation, and tool holders without locking mechanisms, can also impede operations and cause frustration as well as safety issues. Having to read through deep piles of work instructions to locate the right procedures is also a significant waste of time. Although workstation items may be in close proximity, poor organization and station design create unnecessary motion.

Low-Priority Waste

Let's go back to the article I read on my plane ride. It should be obvious now that lean manufacturing is a continuous journey, and that makes it nearly impossible for a company to be 100 percent waste free. Continuous improvement is the name of the game.

When I think about that "100 percent waste free" company, I wonder how it moves its product along the line. What approach has it taken to remove the movement of parts and product? Automated or not, this movement is non-value-added, and it's an example of wasted motion. But it is nearly impossible to remove this movement from the build process. Some companies have implemented continuously moving assembly lines, where the operators walk as they install the parts. I do not recommend this practice, and I believe it is not conducive to good quality and safety. Let me give you an example of what happens when the line moves.

A company that manufactured vinyl windows attempted to install a moving line in one of its high-volume plants, where it operated approximately 22 manual and automated assembly lines. Some lines had conveyor belts, which were operated by the worker at the start of the process. He would finish his work on a window, place the window on the conveyor, and press the foot pedal to move the belt. The product moved, and operators on the line were forced to move with it while performing assembly, no matter where they were in their own assembly process.

It reminded me of an old *I Love Lucy* episode with Ethel and Lucy trying to put chocolate candies into boxes that moved past them at an increasing pace. Soon the two women weren't able to keep up, and they started stuffing the candies into their clothes and then into their mouths. It was funny on TV, but in real life it's anything but amusing. An analysis of internal and external quality data at the window company revealed a significant number of parts missing from the windows that flowed on conveyor belts. Even more telling, the parts most often missing were those that had to be installed. Because installing required time, the operators simply skipped that step; they had no time to do it. I also observed many operators lying on the moving conveyor belt, attempting to slow it down so that they could install parts.

My point is that some waste is nearly impossible to remove, and attempting to do so is not expected or desired. If you focus on the high- and medium-priority waste, you will make major improvements in your manufacturing processes. Don't worry about not having enough to do. Remember that this journey is never-ending. You will be continually faced with inefficiencies and waste that will require removal.

Quality at the Source

With regard to quality, one of the most common mistakes is to implement self-checks and successive checks incorrectly. As mentioned earlier, the quality at the source philosophy places the responsibility for quality at the point of build. Ultimately responsible for product quality, operators must perform certain incoming and outgoing checks throughout the process. They check work done in the preceding process or by a former worker, then perform their own task, and then perform a quality check on the work they just did.

Quality at the source results in a tremendous improvement in quality. When checks are performed throughout the process, multiple eyes are on the product. This practice results in a product that is virtually error free by the time it reaches a more formal inspection point at the end of the line. Self-checks and successive checks are very common in a lean journey, but only a correct implementation of these checks will ensure that they are performed.

Never implement quality at the source without first identifying the current state by conducting time and motion studies. The incoming and outgoing checks require additional minor effort from the operators and add to the lead time of the product. An early stumbling block is to add this new operator requirement without first looking at the current process, resulting in headaches, additional lead time, and lack of accountability.

Here's the right way to do it. First, document the current state of the build sequence: the sequential list of the work actions performed in the process. When you balance the workstations, add the in-line quality checks, and be certain to allocate time for each check so that workstations are not pushed beyond takt time. Eventually, these checks will become part of the standard process, so be sure to allocate the time correctly.

Don't overload the production workers with too much to do. Remember, they are not inspectors. I usually advise my clients to assign no more than four checks, in any combination, to one workstation—for example, two incoming and two outgoing, or one incoming and three outgoing. However, not every workstation requires four checks; some may need only one or two. The maximum should be four per station.

Another mistake in implementing quality at the source is the type of checks that are required. What should the production workers look for in regard to quality? There are a multitude of items that can be checked and verified during the assembly of a product. Checking criteria should depend on your product, processes, and customer requirements.

It is important to select the checks using quality information, which can be gained from various groups within the company that track external and internal quality. Review the data derived from internal inspections. Quality information can also come from customer complaints and issues raised by outside service technicians. For example, perhaps there has been a continual problem with a wire harness not being fully seated in its terminal. The customer's product may not be performing as expected, and the internal quality technician has had issues with the wire harness as well. It appears that the problem is a common one and needs to be addressed. In this case, checks will be initiated based on real data and not simply on opinion or "feel."

Self-checks and successive checks can result in dramatic improvements in quality. By checking the major issues, a company can ensure that quality products are being built.

Workstation Design

When you attempt to apply all the appropriate lean tools to a new workstation or manufacturing line, you can make minor mistakes that can negatively affect performance. For example, managers are often drawn to trade catalogs that promote lean workstations and lean materials racks. Although these are possible solutions to line flow and materials presentation, before investing in physical modifications each company should look carefully at its own processes and decide how the most efficient line will operate.

The rest of this chapter is devoted to explaining common mistakes in workstation and line design and offering guidelines on how to choose the best solutions for your processes. You need to customize your processes to fit your specific needs rather than relying on a solution from a catalog. Here are the issues to consider:

- Lines versus work cells

- Physical flow

- Materials presentation

- Tools presentation

- Operator fatigue and safety

- Painting and lighting

- Documentation

Lines versus Work Cells

The concept of work cells has increased in popularity over the years. **Work cells** are essentially teams that work together as autonomous groups, each building a specific line of products. Each work team is responsible for the quality, volume, and productivity of the entire cell. Manufacturing companies are transforming traditional assembly lines into work cells as part of their lean journey. There have been successes, as well as failures, with this transformation.

It is important to note that work cells are not appropriate for all manufacturing environments. They were initially created for small groups of people building small products using a minimal number of parts. The teams in a work cell work very close to each other, and there are very few parts and materials in the area. Operators can move from workstation to workstation, helping the team with flow and progress; therefore, work cell operators need to be highly skilled and cross-trained.

To reduce the number of parts in the process, materials are replenished more frequently in a work cell than on a line. I have seen work cells in which parts were replenished every 30 minutes. It was a highly organized system with aggressive 5S and visual management. The company had worked with its suppliers to bring in low quantities of material and had streamlined its internal inventory control and materials handling to accommodate frequent replenishments.

Companies that operate processes with large, bulky products and parts may not be successful with work cells. Work cells are optimal for smaller or highly customized products. Many organizations attempt to use work cells for long, multistation lines, but that makes the cells long and wide and defeats the concept of work teams and high visibility. Operators cannot easily move to and from workstations, and the large products require large parts, necessitating space for storage. Although this arrangement is not optimal, I am not implying that a long, multistation line cannot be broken into smaller cells that feed into each other. It can be done. A company needs to evaluate its specific process and decide whether work cells will be efficient. The same principle applies to materials replenishment. Large, bulky parts may not be delivered as often due to physical strain.

Be careful not to buy in to work cells instantly. You must stay within certain variables to make work cells operate effectively. Evaluate your processes and determine whether they are good candidates. Remember, lines can still be lean and are perfectly suitable for most manual and automated assembly processes.

Physical Flow

How does your product flow? How will it flow after you implement lean manufacturing? These questions must be addressed. Products must move from one workstation or one process to another. As I've mentioned, this physical movement is wasted motion and you must apply careful consideration when deciding product flow. What works well for one manufacturer may not be the best solution for another. I will present a few options and then discuss the pros and cons of each.

Conveyor Belts

Conveyor belts are very common. Automated or manual, a conveyor belt allows the product to move along the line with very little effort, thereby reducing physical strain. In my opinion, this is a conveyor belt's only positive attribute. Automated conveyor belts are typically controlled by foot pedal at either the beginning or the end of the line. If the operator at the beginning of the line controls the foot pedal, it can create an early push. Once an operator has completed her work, the unit is pushed to the next worker. Ready or not, here it comes! The remaining workers are then forced to work while the product is in motion. On the other hand, if the last operator is in control, there could be an early pull, and although a pull system is the ideal condition, a pull signal can be initiated improperly.

Again, operators are working with a moving product. From an ergonomic perspective, operators are confined to one specified height that is set for the team as a whole.

If you choose to use conveyor belts, you will need to deal with culture change early in the process. The best approach is to ensure that operators flex within the workstations, a practice that allows all operators on the line to perform work within the specified task time. When each worker has completed his work, the foot pedal can be pressed to move the belt for the team and not the individual.

Conveyor Rollers

Conveyor rollers are very similar in use to conveyor belts; both move product. However, belts control all the workstations on a line, whereas rollers physically divide the workers and do not move the entire line. Automated conveyor rolling systems can be implemented, and they are essentially like belts. My focus is on manual roller systems.

Roller systems are built to a predetermined height. However, the rollers can be modified to fit individual workstations or lift tables. Operators pull units into their workstations as needed. But, as with belt systems, an early push or pull can be forced. Use of roller systems requires training and then control to ensure that products are not forced onto other workers. Conveyor rollers work well when used in conjunction with other workstation items, such as benches and lift tables.

Lift Tables

Lift tables are one of my favorite choices if the target process or assembly line can use them effectively. Operators are not confined to a predetermined height and can adjust and maneuver the unit to a height that is suitable to their personal working requirements. I call this "climbing in the product." Lift tables come with a foot pedal that is used to raise or lower the table to the desired height. Lift tables come in a variety of sizes, so there is one to fit every type of line.

As I've mentioned, adding conveyor rollers to a lift table provides more options in product flow. Production workers can place the unit at a productive height and also can easily pull units into their workstations. When you balance the line, be sure to account for the time associated with up and down motions of the lift table. Although it may take only a few seconds, when multiplied over numerous workstations throughout the day, this time can add up. Of course, movement time is wasted motion, but I consider it low priority because it is a valuable addition to the process.

Workbenches

If you are opting to use workbenches in a manufacturing line, make sure you analyze the need for them. From a materials and tools presentation perspective, workbenches are great because parts and tools can be placed directly in front of the worker. Adjustable workbenches are available that have a lever that can be turned to raise or lower the table.

When you analyze the need for workbenches, take into account that they often attract chairs or stools. I have never been a fan of chairs and stools on the production floor, because they take too much space and can place operators in poor sitting positions, causing back and neck strain. Does the work warrant the need for chairs? Small products require intricate work, perhaps with a magnifying glass, so these production lines may be good candidates for chair use.

On most other lines, production workers should be standing in front of product at all times. Standing encourages attentiveness and increased productivity, whereas chairs and stools may cause operators to become complacent or to slow down. If the workbenches are set at waist level in the station, you can avoid the use of chairs altogether. However, it is important to remember that workers may be standing for eight hours at a time. I discuss operator fatigue later in this chapter.

Mobile Lines

Assembly lines that can be moved easily, or rolled around the factory floor, are very handy. I once worked for a company that used mobile lines. They allowed flexibility, because the company could maneuver the processes as needed. Some products are heavy and bulky, requiring a mobile approach for efficient product flow.

Pallet jacks, pallets, carts, and so on are only a few types of mobile units that can be used on a line based on product requirements. For example, if your products require thick steel casings, as do safes or ATMs, placing them on pallets and moving them with pallet jacks is the recommended solution. Although there is some manual movement required within the process, you greatly reduce the possibility of injured workers or the product falling and becoming damaged.

I assisted a company that removed an old conveyor roller system that had been used to build medium-sized engines. The roller system was replaced with customized carts, each with a small platform on which the engine was secured during assembly and transportation. These mobile carts were height adjustable and moved through the assembly line like a train. The wheels were also lockable to avoid uncontrolled rolling of the carts. When the engine was lifted off the cart into the next process for installation, the empty cart was returned to the beginning of the line. The materials and tools racks were also on wheels, so the entire line could quickly be rolled out of the way for cleanup or inventory counts.

Material Presentation

Parts and materials are the blood supply of the manufacturing process. Therefore, they must be presented to operators in the most efficient manner possible. Manufacturers must remember the most fundamental aspect of materials presentation: point of use. With point of use comes the need for an effective materials replenishment system.

Materials handlers are a vital part of this system. In *Lean Assembly: The Nuts and Bolts of Making Assembly Operations Flow* (New York: Productivity Press, 2002), Michael Baudin compares the importance of a materials handler to the importance of a race car pit crew to its driver. In his example, the driver represents the line operator, racing around the track, jockeying for position. Major focus is placed on the driver as the value-added person on the team. However, at some point the driver must make a pit stop and refuel, get new tires, get clean windows, and receive water. The people in the pit crew are the materials handlers, servicing the car and driver before the driver heads back onto the track. Crew members follow clear, established procedures in a specific sequence. No matter how efficient the driver is, if the pit crew is not successful on every pit stop, the race could be lost. It is a collaborative effort that requires successful work on both ends.

When materials are presented to the operators, the materials must be ready for installation. As mentioned previously, packaging should be removed during the receiving stage before workstation presentation.

Each workstation requires a method for replenishment and a form of clear communication and signaling. For example, if the line is using the two-bin system, in which one empty bin is the signal for more parts, the established quantity must be identified either on the bin label or through the use of tower lights. Operators and materials handlers should be in constant communication, and material must be placed in the workstations in minimum quantities. Remember, the line is not a stockroom. Note that the two-bin system is outlined in detail in my book *Kaizen Assembly: Designing, Constructing, and Managing a Lean Assembly Line* (Boca Raton, FL: Taylor and Francis Group, 2006).

Tool Presentation

There are basically two approaches to tool placement: waist height or overhead. Tools should be easily accessible and never placed on the floor.

The best placement for air hoses is overhead. If the air tool must be presented at waist height because of the work performed in the workstation, be sure that the hoses are routed along the work area and kept off the floor. All hand tools should be placed on shadow boards (in which tools are depicted or outlined to show where they belong) or on a tool trolley. Tools should be clearly identified with a label that specifies all critical information, including appropriate locations and quantity for the specific workstation. Tools should never be shared among operators.

Be sure to develop and sustain a tool maintenance and management program, which will ensure that tools are cleaned and in proper working order. Always have backup tools on hand in the storeroom, and keep track of calibration, torque setting schedules, and time associated with repairing and replacing tools. This practice will help you to identify trends in trouble areas and will contribute to improvement efforts.

Operator Fatigue and Safety

Another common mistake made in process design is failing to account for the fatigue experienced by operators when they stand on concrete for long periods. The use of antifatigue mats is an easily adopted measure that can increase productivity and improve operator morale. When selecting a mat, choose a style that cushions the foot. Test it by standing on the mat. A good mat shows a footprint after your foot is removed, meaning that the mat has absorbed weight and removed some fatigue. Create a mat replacement system to ensure that operators always stand on mats in good condition. Remember, antifatigue mats eventually experience fatigue, just like everything else. Replace them when their product life cycle is over.

During factory tours, I have noted potential safety issues related to improper use or nonuse of regulation safety glasses, earplugs, and gloves. A safety analysis is a must! Protective eyewear may not be needed plantwide, but certain operations that involve rivets or nails warrant the need for eye protection. Earplugs may also be necessary, depending on noise and decibel level. I recommend protecting people as much as possible.

Painting and Lighting

Manufacturing factories should be bright and have a pleasant appearance. Painting the production floor is always recommended, although it can be costly. Manufacturers that have painted their factory floors have found added benefits in morale, productivity, and quality.

Workstations and work areas should be well lit. Do not rely on the fixtures hanging from the factory ceiling to provide adequate light. Many factories have a hazy appearance because the lighting is placed too high above the lines.

Bright lights and bright paint encourage a sense of performance and a good disposition. If you walk into any fitness center or gym, you will see that they are well lit and cheerily bright.

Documentation

Like parts and tools, work instructions need to be available at the point of use. Work instructions should be placed in each station and not hidden in a cabinet.

Each work instruction should be developed based on the work performed in each target station. Unfortunately, documentation is sometimes a last-minute effort for busy engineers. Often, they simply copy and paste drawings or pictures from other instructions in an attempt to complete the documentation quickly. But the pictures are not always suitable for the work being performed. They may show the product without any parts installed, with parts from a prior workstation installed, or with all parts installed—but with no instructions on how the workstation should perform its specific task. Using inappropriate pictures or drawings only creates confusion for the operator.

Work instructions should always be a step-by-step process of how to complete a specific task within a specific workstation. That's all—nothing more, nothing less.

Chapter Wrap-Up

Some of the information contained in this chapter may seem just a matter of common sense. But curiously enough, many of the simple approaches I've described are often ignored. Organizations tend to repeat the most common mistakes I've discussed in regard to data collection, 5S, and process design. Attention to details can make a big difference in your lean journey. Everything you choose to do must be a good fit for how your company operates. Take each suggestion as needed, and lean it out.

five

Operator and Supervisor Involvement

Lean manufacturing has been praised as the saving grace that can bring a business to world-class status. But the truth is that it does not come easy. As a consultant, I try to guide my clients along the path of least resistance, but resistance comes at every level. Operators and production supervisors enjoy their established routines and don't want to change them.

We all tend to resist change. For example, commuters establish certain driving patterns to and from work, and typically they follow the same route every day unless they have errands. Drivers often choose a highway as their regular route because of its higher speed limit. If there is an accident and traffic begins to back up, drivers realize that it will take longer than usual to get to their destination. Perhaps the police will redirect traffic toward a detour route that is longer and has a lower speed limit. The delay causes drivers to become unhappy because they are experiencing something different from the norm. They are resistant to the change in their established routines.

In another example, let's say you have been driving the same route for years, and, over time, the road has deteriorated, with scattered potholes and faded, worn-out road signs. Your commute has become increasingly complicated and you begin to complain about it. You may think, "Why doesn't the city fix this road?" And then, suddenly the road is approved for an upgrade and repair. Now you are confronted with traffic workers signaling you to slow down or stop. Road workers and equipment are spread out all along the road. During this time of improvement, there will be many obstacles in the road and challenges to your schedule. Although you realize that the repairs are badly needed, it still makes you unhappy.

A few weeks later, as you drive home from work, you notice that all the workers are gone, and the road work is completed. You are delighted with the improvements, and you can see that they will make your daily commute much more productive and enjoyable. Suddenly, you are no longer unhappy and no longer complaining. You recognize the value in having to go through the pain to get to the gain.

Welcome to your lean journey! The behaviors outlined in these examples, and the changes the drivers undergo, are very similar to the kinds of behaviors that people experience during a transition to lean manufacturing. It's normal for human beings to be resistant to change, especially when the change first occurs and they cannot see the benefit.

Standard Work

One of the fundamental principles of lean manufacturing is the concept of **standard work.** Standard work is the best, most reliable, and safest way of performing work, and it drives the protocol in a reduced-waste environment. Standard work calls for clearly defined roles and responsibilities for every worker, supervisor, workstation, and piece of equipment.

Before you can implement standard work, you must eliminate or reduce waste and inefficiencies, such as overproduction, overprocessing, and wasted motion, transportation, inventory, and waiting. Standard work can then be set in place and used to monitor and control a lean process. Of course, you will always continue to identify ways to improve standard work as part of your ongoing lean journey.

For the lean process to be successful, operators and production supervisors must follow the standard work that has been set in place. The transition process is as follows: identify waste, remove it, create standard work, and follow standard work. Standard work defines all the work that must be performed in a process and the rules that govern that work. Here some examples of standard work:

- Established work in a station

- Established quality checks in a station

- Build sequence (the order of work to build a product)

- Materials handling routines

- Equipment start-up tasks

- Machine setup steps

- End-of-day cleanup

- Forklift routes

- Work instructions

- Equipment use instructions

- Testing and inspection requirements

Standard work should be created for every job and task needed to run a specific process, with each job clearly defined and supported by data and documentation. Operators will find working within standard work to be very rewarding. It is easier to learn operations when everything is perfectly outlined. There is no confusion about which work is performed in which workstation, which route the materials handler takes, or how to start a piece of machinery. All tasks are clearly specified from the beginning. Clear direction allows the operator to move easily between stations and perform the work as needed. When all the work is clearly defined, operators can catch errors and problems in the process because any deviations from standard work are noticeable. Standard work allows the operator to identify and suggest needed improvements. It is a simple approach to work.

The company also benefits from standard work. Standard work reduces variability in the process, lowers costs, reduces waste, improves quality, and helps establish shorter, more efficient lead times for customers. Problems are quickly visible and the company can adjust as needed to bring the process back into control.

The best part about standard work is that it is result driven and measurable. Removing waste and creating standard work allow a company to design and implement efficient manufacturing processes, and that, in turn, allows for more predictable lead times, clearer staffing requirements, increased quality, and reduced cost. When standard work is implemented in a work area, the area can operate effectively with minimal effort.

Standard work environments lend themselves to improved performance and control if operators work within the standard work rules. Reverting to older ways of working in a new environment will only create problems. It is understandable that workers will initially resist the change, but at some point they must adhere to the new standards. When assembly lines or other manufacturing processes contain a great deal of waste, there is a lessened sense of urgency; people simply work at their own pace, which can be too fast or too slow. Operators can build up excessive WIP (work in process) and then take extra breaks. Workers can return from breaks individually rather than as a team. Also, they may leave their workstations to find tools and parts because the line was not set up correctly.

Production supervisors must accept the responsibility to keep people accountable if they fail to follow the standard work. This is a difficult role, because it is all about dealing with people. Operators tend to resist change at first, but supervisors resist change the longest. Before the improvements, they could work at whatever pace felt comfortable. Now they are required to work at a quicker pace while following the rules of the process. Production supervisors may try to revert to old ways of supervising, but it is impossible to lead in the same way as before. It just won't work. That is why the relationship between operators and supervisors must be strong.

Following are elements of standard work that must be followed, and what is expected of the supervisor and operators in a lean process:

- Following the work content
- Using single piece flow and controlled batches
- Staying in the workstation
- Maintaining communication
- Working within effective hours

Following the Work Content

During a lean journey, manufacturing lines are designed around takt time. **Takt time** represents pulse or rhythm and is established by dividing the amount of time available to work by the number of units required from a given process. Takt time is used to design a manufacturing line so that a certain number of products can be built at a given interval. It represents the completion time of an interval. For example, if takt time is 5.5 minutes in a given assembly process, then a product must be completed every 5.5 minutes. Engineers and kaizen teams then balance the work content so that every workstation is as close to 5.5 minutes as possible.

Following this set work content is critical, especially when single piece flow has also been implemented. Units in the line must move every takt time to ensure that the required output is obtained. Operators must follow the work outlined in the work instructions, and production supervisors must enforce this practice.

Deviations from the work in the workstations is highly noticeable. Possible bottlenecks may appear. Buildup of WIP or excessive waiting are clear indications that operators are not performing the required work. Unless there is a problem with parts or there is a quality issue, operators must follow the standard work in their stations.

In the beginning of a lean journey and as lines are improved, failing to follow standard work will be one of the initial attempts at resistance. Production supervisors must show confidence and strength and must react to any imbalances quickly and firmly, something that is harder than it appears. Operators tend to migrate to the type of work they are comfortable with, and there is a good chance that work will be distributed differently after waste reduction and line balancing.

The standard work content in each workstation is established for good reasons. There is also an established time standard for this work. During time and motion studies, time standards were captured to help with the design of the lean process. Operators must be trained to this standard, and then they must be monitored by the production supervisor to ensure compliance. It is also important to instill a sense of urgency in the operators. Manufacturing lines should be designed to work at an efficient pace.

The concept of **operator loading** comes into play here. Never load people to a 100 percent pace, because they will get tend to become stressed and frustrated. In addition, 100 percent loading of people is not conducive to good quality and safety. Thus, you should be sure to add a small percentage of personnel, fatigue, and delay to the line's design volume, time standards, or workstation cycle times. Even with this buffer, the pace of the line should be productive. Maintaining this pace is critical to the success of the line. Waste within a process allows people to slow down as they see fit and perhaps work at a pace that is not optimum for the manufacturing environment.

I once discussed pace with a manufacturing plant manager, and he agreed that people should not be overloaded. He also commented, "They are not knitting, either." His point—and I agree—is that the factory floor should be moving at a productive pace that promotes efficiency, quality, and safety.

Using Single Piece Flow and Controlled Batches

Operators and supervisors also need to understand how to work in an environment with reduced WIP. This can be very difficult. Single piece flow is a lean concept: working on one unit at a time, with either no units (or at most one unit) between workstations. If operators and supervisors are not familiar with this type of environment and it has recently replaced a different kind of process, there will be an adjustment phase. Single piece flow is usually associated with takt-time-driven processes. The combination of the two concepts creates an even higher level of urgency. Units are now flowing through the process very quickly, because piles of WIP no longer provide a buffer.

During this transition, as operators and supervisors adjust to the change, the line is not likely to achieve its desired volume. Single piece flow processes also require operators to work as a team, a concept that most likely is new for them. The assembly line will not be perfectly balanced, especially at first, and will require some adjustments. It is difficult to set workstations to exactly the same cycle time, but you can get fairly close. Optimum balance is challenging, and teamwork is the key.

Operators tend to become personally connected to their workstations and to the work they have performed for several years. In a non-single-piece-flow environment, operators typically perform only their own work. Piles of WIP create a sense of isolation between stations and processes. When single piece flow is implemented, you cannot expect the operators to transition instantly from individual workers to team players. This is not necessarily negative; it is only human nature. However, the old culture will have to be remolded into a new culture of lean workers who perform as a team to help keep product flowing.

The imbalances I've referred to may be approximately 5 to 15 seconds between workstations. An old solution would be for the operator to start working on another unit and place the completed unit near the next worker or on the floor. This is overproduction. The operator may be attempting to stay busy, but this action does not add value to overall flow.

Supervisors must react to this potential bottleneck. Operators must flex back and forth between stations, helping out as needed. Here is a simple rule: Never take work with you to another workstation. The work content in each station is for that particular station only. If production workers are to move, they must help do the work defined in that workstation. This practice allows the product to move smoothly and efficiently throughout the process without major stops or slowdowns.

Single piece flow is established in order to reduce overproduction and surplus inventory. As you can see, there is a different method of working in this type of process, and managers and supervisors must be skilled in dealing with the newly transformed work culture.

Be aware, however, that single piece flow is not 100 percent applicable to all companies and processes. For example, you cannot apply the concept to a job shop environment, such as machine shops, or to organizations with fabrication-style equipment. I would not expect a brake press operator to set the brake press for a run of only one unit. In this case, a different methodology must be established that meets the needs of the production line.

Controlled batches is another form of flow control. In a **controlled batch,** the products (or parts) are fabricated or assembled to a set quantity. Once that occurs, the operator is required to stop, check the work, and then move to another product in need of processing. As with single piece flow, a learning curve exists when this process is implemented. Although controlled batches allow the operators to work on more than a single piece, they must learn to stop at the set quantity and move to the next item required.

Controlled batches are often used in subassembly work cells, where there are a number of operators building a variety of assemblies for other consuming processes. They must work as a team and move to and from subassemblies, building to the established quantity. Management must properly cross-train employees to be able to perform all the types of work. Having an operator who works only on the subassembly process helps the overall process. The idea is to build as needed, at the right time, and in the right quantities. Production supervisors must learn how to respond when the build quantities are not being completed properly.

The best approach is to have the build quantities displayed visually on signs in the workstations that specify the item and the required quantity. The production supervisor is required to walk by and observe the finished goods area of the work cell. Using this method, the supervisor can quickly see whether the process is under control. When subassemblies are not being built as required or are not in the correct quantities, it is very noticeable and indicates a potential problem.

Staying in the Workstation

This is a matter of common sense, right? Easy to control? Guess again. Operators are value-added employees. Over the years, some managers have grown tired of this definition, but it is still a fact. Employees who manufacture and assemble products are creating revenue for the business. Sometimes referred to as **direct labor,** operators build product, and that allows everyone to be paid.

That being said, it is important that operators build as much product as optimal for the company. For this to happen, manufacturing processes should be designed to keep operators in their workstations. Work areas must have everything operators need so that they do not need to leave their stations.

I want to emphasize, as I have before, that operators should not leave workstations for unrelated work activities or personal reasons. Taking extra breaks, leaving to talk on cell phones, going to other lines to talk, or using any other reason to leave a workstation (except for official work breaks, lunch breaks, required meetings, and the like) must be prohibited.

For most organizations, this practice requires a culture change. Production workers, as well as supervisors, tend to do as they please while working. What's more, relationships often develop between supervisors and operators over time; after all, most supervisors are former line operators who've been promoted. Managers want to encourage internal advancement because it is healthy for the company, but the buddy relationships that may exist between former and current line operators may be a stumbling block to enforcing the rule.

I am not suggesting that you create a labor camp at your factory, but the lean process requires a specific method of working, and initially that may be difficult for managers. Everything is about performance and following the rules of the process. We want production workers to stay devoted to their workstations to ensure output, quality, and a sense of teamwork. Production supervisors cannot simply let operators leave workstations whenever they want to, so supervisors must be committed to the new process, providing a good model for the workers.

Maintaining Communication

Lack of communication—from the top of the enterprise all the way down to the factory floor—can bring a lean journey to its knees. Establishing and maintaining an effective communication system between operators and production supervisors will help ensure that the process operates correctly.

The key factor in communication is to make sure that operators can find a production supervisor whenever they need one. The process should be designed to allow operators to continue working after signaling for help. As just mentioned, the goal is to keep operators in their workstations while they contact their supervisors. From my experience, I know that production supervisors are very busy people, and they may not always be readily available. One solution is to install communication lights.

Communication lights, often called **tower lights,** are installed in a workstation to act as communication signals from operators and supervisors. Communication lights come in a variety of styles and colors. I have found that the best type are those that have red, yellow, and green lights, similar to traffic signals. Having these three colors supports optimal communication. You can define the meaning of each color based on your needs; however, here are my recommendations:

- **Red** means that there is a major problem. It indicates that immediate assistance is required; the problem is large, out of control, and must be addressed. Typically, the red light is lit when a tool breaks down unexpectedly, a machine has stopped, there is a severe quality problem with a part or unit, or an injury has occurred to a worker.

- **Yellow** means that a problem is coming but is not a major one. You want to instill proactive thinking in your employees. Potential problems should be caught before they become major issues, and the yellow light should be used to communicate the potential problem to the production supervisor. It is a warning. For example, a yellow light might be used when material is running low, a tool appears to be working improperly, a machine is making an odd sound, or an operator needs a bathroom break. In the case of the latter, the supervisor must fill that position until the operator returns. The line should not be halted for a bathroom break.

- **Green** means that there are no problems. If the process is in control, product is flowing, there are no quality issues, and material is being replenished, all communication lights should be green. A production supervisor can walk out to the factory floor and very quickly see that the lines are performing as expected.

This communication system presents a learning curve. Operators must learn to use the lights properly to ensure that communication is open and honest regarding problems occurring in the work area. The lights should never be used in error or with disregard for their meaning, because false signals cause unnecessary confusion, attention, and distraction and negatively affect the work process.

Effective communication must also take place with regard to the lean journey. Supervisors need to alert their operators to upcoming kaizen events and let them know when the engineer or kaizen champion will be conducting an analysis. Data collection is an absolute requirement for lean manufacturing, as discussed in Chapter 4. Time studies and other analysis or data collection activities can be stressful for production workers.

I have watched many companies as they begin their continuous improvement initiatives. Often, engineers and technicians hit the factory floor with stopwatches, cameras, and clipboards without any prior warning. That is not recommended. The company must involve the production supervisors from the beginning, informing them about the team of data collectors coming their way.

Adequate time should be allotted to allow the supervisors and engineers to talk with operators about the analysis that will take place. Operators will be asked questions during the analysis, so they should know about it ahead of time to lessen anxiety and to encourage a positive attitude and input. If you have bought in to the company kaizen program, as outlined in Chapter 3, then you know that these operators will play an active role in the kaizen events and in contributing input to the changes. You want to ensure that effective communication takes place early in the process so that you get everyone in the loop, and on board, in a positive and supportive way.

Working within Effective Hours

One of the highest costs in a factory is labor, and focus is often placed on the operators and their performance within an eight-hour period. If operators are being paid to work eight hours a day, the thinking goes, we should measure their performance for eight hours. Unfortunately, this is not an effective approach.

Instead, companies that are seeking to implement lean manufacturing should understand the concept of **effective hours** (EH), which reflects the amount of time that operators are actually working with product. Operators do not build product all eight hours of their shift. Although they may be in the building for eight hours, they do not spend the total eight hours touching the product. I called the time touching the product **touch time.** Here is an example of how a typical eight-hour shift is spent:

Time in the building	480 minutes
Morning meeting	15 minutes
Morning break	10 minutes
Lunch (paid or unpaid)	30 minutes
Afternoon break	10 minutes
End-of-day cleanup	15 minutes

Of the total 480 minutes that the operators are in the factory, 80 minutes is spent in activities not related to working on product. That leaves 400 minutes (approximately 6.67 hours) as effective work time. The calculation for effective hours will vary. Some organizations operate under four 10-hour days, so the final effective hours will be different. In any case, you need to establish the time associated with building products and fabricating parts.

From a management perspective, the operators are getting paid for 8 hours of work, but, based on the example, only 400 minutes, or approximately 6 hours and 40 minutes, is actually work time. This time must be used productively and not spent leaving the workstation, looking for parts and tools, performing long setups, and searching for production supervisors. If 20 percent of the time is spent in wasted movements and communication, the effective hours, or touch time, will be only 5 hours and 20 minutes. Do you see my point? Organizations cannot afford to have operators deviate from the standards and rules of the lean process. It is up to the company to make continuous improvement a major focus in regard to effective hours. As a rule, to get consensus you need only provide a sound explanation of the importance of continuous improvement so that the company can use workers' touch time effectively.

Operators and production supervisors need to understand the concept of effective hours and then learn how to work within the desired time frame. It is critical that operators work as team at all times and leave and return from breaks together. Production supervisors must manage their people accordingly to ensure that they are maximizing their time. Conformance won't be instantaneous but will take time and practice as well as supervised monitoring.

Not working within the established effective hours can result in the need to work overtime, which is usually viewed as a negative effect. The thought of working late or over the weekend will quickly convince the operators of the importance of making effective use of their time. Of course, it is up to the company as a whole to create an efficient work environment and continually drive waste out of the manufacturing processes. Once that is accomplished, operators and production supervisors need to follow what is in place for the process to run smoothly. It is a total team effort.

How to Get Them Involved

So far in this chapter I have discussed the importance of the relationship between operators and their production supervisors, and I've explained how effective interaction between them will ensure a successful lean process. How do you get them involved and following the rules of a lean process? In other words, how do you change the floor culture? Mistakes are made in this area again and again, and there is no perfect guideline to follow. However, I'll provide a few ideas to help reduce the resistance—reduce being the key word.

Using the Kaizen Suggestion Box

Chapter 3 is dedicated to establishing the company kaizen program. There is a tremendous amount of detail in establishing that program, and I would like to tie some of the elements to this chapter.

The kaizen program, as a whole, is a great way to develop a sense of continuous improvement and involvement. One of the items discussed in Chapter 3 is the employee suggestion box. The key to a successful lean journey is buy in, and it starts with communication. The suggestion box is for use only by the production workers. As a proponent of a lean organization, you want the suggestions for improvement to come from the factory floor. Operators can submit ideas for improvements to their own areas or to other processes.

As simple as it may appear, it works beautifully. In my experience, there has never been an empty box. Operators appreciate the fact that their ideas mean something to the company. Of course, you want to encourage suggestions face-to-face, but some people prefer a less personal approach. The suggestion box gives them that alternative.

After the kaizen champion and the kaizen steering committee review the suggestions, they will decide which ideas will be used during the next kaizen event. Ideally, the operator who made the suggestion will be placed on the kaizen team so that the employee can help turn her ideas into reality. Again, this is ongoing involvement, and the beauty is that it works. As you conduct more kaizen events, more operators and supervisors will become involved in the implementations.

To gain buy in from operators and supervisors, it is important to involve them in kaizen events. It is good for the culture to see everyone contributing to improvements. After the kaizen event, the operators and supervisors can help change the culture in the new lean process. Because their ideas were encouraged and implemented, the resistance on the line should be reduced. They can help train and mold the other line workers into positive change agents.

Kaizen Steering Committee Floor Representatives

Over the years, I have taught companies the importance of establishing a kaizen steering committee. As you can recall from Chapter 3, I discussed the key managers who should sit on this committee. These included the upper and middle managers, who manage various departments in the company. Basically, they are the key decision makers for the organization and should help to plan all kaizen event related activities. One way to get involvement and "buy in" from the operators and supervisors is to appoint a representative from the production floor as a committee member. I did not mention this option in Chapter 3, as I felt it pertained more to this current chapter.

Having a floor representative on the kaizen steering committee will help reduce resistance on the floor. With a worker from their area on the committee, the production workers and supervisors feel confident that someone is representing their specific needs at the monthly kaizen meeting. The input of the floor personnel is just as important as the input of the other committee members. When the committee discusses which areas of the factory will be scheduled for a kaizen event, the floor representative can give input from his perspective and can help decide which floor workers should be part of the kaizen team. In a union environment, the best candidate for floor representative is the shop or union steward.

Involving Operators in Data Collection

As I've mentioned, lean manufacturing is data driven, and you need to collect a significant amount of information in order to make effective improvements to the floor. I discuss some of the waste analysis tools in Chapter 4 and mention the importance of involving the operators. The kaizen champion will consistently analyze the processes and gather valuable data on the current state. He will conduct time studies, waste analysis, possible process mapping, and other exercises. This data collection is ongoing, in preparation for future kaizen events. Although the kaizen champion is dedicated to the majority of the lean initiatives, he must involve the operators and production supervisors during data collection.

The kaizen champion uses this information to come up with improvement ideas and possible changes in the line. Operators and production supervisors should be consulted, on an ongoing basis, to provide assistance to the kaizen champion. It is wise to schedule time after data analysis for the key operators and supervisors to meet with the kaizen champion and give their feedback. They might identify a line balancing error, work content mistakes, or missing information. Perhaps the kaizen champion has evaluated an area and misidentified the root cause of the problem. In this way, operators and production supervisors can help reduce mistakes early in data collection and analysis, allowing the kaizen champion to continue with correct information. This arrangement also allows operators and supervisors to become involved at the very beginning of the lean journey.

Lean Manufacturing Training

Training is key to a successful lean journey. Generally, an organization sends engineers and managers to some form of lean manufacturing training. Engineers and managers are typically the individuals who will drive the improvements and plan the events, and therefore they need the training. However, production workers and their direct supervisors also need training. To ensure consistency in learned methodologies, all employees should go through the same lean training.

I realize it is difficult to schedule a full day, or even a half day, of training for production workers, because they are performing value-added work. Nevertheless, some form of lean manufacturing training is critical. In one situation, I was hired to train a start-up manufacturing company in Burlington, Washington. It had hired me through Skagit Valley College in Mount Vernon. My charter was to train every employee on the concepts of lean manufacturing. I was contracted about three months before the expected date of start-up.

Three groups of employees required training—the line leads, the operators, and the support staff—and the training was spread out over a couple of months. The company had decided to train its employees in additional skills as well. Although in start-up mode, the company felt that all production workers would benefit from an intensive week-long series of courses. Company managers realized that some information would be forgotten, but they saw a window of opportunity to provide sufficient training and decided to take it. I was impressed by their foresight, because it demonstrated the dedication they felt to their employees.

The line leads were trained first, followed by the line operators. I liked the fact that the leads and operators received the intensive training before the managers did. This approach was smart, because the floor personnel were already anticipating whether and how lean manufacturing would affect them. It was a positive approach even though there is a limit to how much implementation can be done during start-up. All the employees attended the same courses, so everyone was on the same page. Management training was done after all the leads and operators had gone through the courses.

Involving the operators and production supervisors early in your lean journey will help reduce the confusion, as well as the initial resistance, that generally accompanies a move to lean manufacturing. Companies fail when they attempt to push lean manufacturing on to the floor culture. Although there is no perfect moment to introduce the concept of continuous improvement, how you involve floor personnel is critical to your success. Managers, too, need to undergo a change in cultural attitude so that they can be good role models for the effort. The companies that do not involve floor personnel as recommended often meet with failure. If you tend to impose changes from the top down, it's time to change your mind-set.

Chapter Wrap-Up

I hope this chapter has provided some insight about the importance of involving operators and production supervisors in your lean journey. As I have stated, often there is no perfect template for reducing resistance, easing culture change, and encouraging buy in. This is why kaizen is called continuous improvement; the process never ends. Remember, the relationship between operators and production supervisors is critical to effective implementation. As a team, they must be strong and committed to the lean process and its success. It is up to management to put the right kind of infrastructure in place to provide adequate training, which ensures that operators and supervisors achieve success as a team and as an organization.

Lean Training Programs

Training new and existing employees takes on a whole new meaning when an organization embarks on a lean journey. Often, companies do not train or invest in their employees in general, and that can have a negative impact on the company's overall performance. When you take a lean journey, you must formally train new and existing employees to ensure that they understand lean manufacturing and know how it is being applied in their factory.

As you implement lean principles and areas of the factory are working under the controlled conditions of 5S and standard work, training helps workers adjust to their new environment. All new employees, including newly hired operators, supervisors, engineers, and managers, must go through a similar process.

Training Programs for New Employees

As your lean journey moves along, it is important to hire people who can contribute to the continuing success of the journey. Their contribution is contingent on how well they are trained. It is wise to introduce lean manufacturing to new employees very quickly and with a lot of excitement so that they get on board right away and realize that lean is a part of how the company operates. Lean is a way of working. Some organizations, when searching for new talent, advertise that they practice lean. It can be a good marketing approach to attract the best employees.

Training new employees requires structure. I recommend that you provide four levels of training in addition to the typical HR new-employee orientation. How this training is structured and scheduled will vary from one company to the next. I will outline the fundamental aspects of the lean training plan, and you can adjust it as needed.

- Level 1: Company product overview
- Level 2: Quality overview
- Level 3: Introduction to lean manufacturing
- Level 4: Mock line training

It takes time to set up the four levels of training and put all employees through the curriculum. Completing mock line training is likely to take at least five days, a period that may be too long (or too short), depending on your manufacturing environment and how you want to conduct the training. You may not be able to afford to train new employees in the recommended four levels. Decisions will need to be made. But regardless of the company restrictions, training new operators is extremely important to the success of the lean implementation and the company as a whole. Consider it something you can't live without.

New managers and engineers often receive formal training when they join a company, and even manufacturing companies tend to invest more heavily in training support staff than in training production workers. Some managers argue that production workers will take their training to any other company that will pay them one dollar more per hour. Simply put, these managers believe that investing in training for line workers will only benefit someone else. In fact, though, it is bad practice to automatically omit training because of the fear of turnover. Many employees will stay with the company, and their training is not a waste of money. If you invest in your people, you will reap the benefits.

Level 1: Company Product Overview

Most organizations have new-employee orientation that involves a variety of topics in regard to human resource functions. Employees fill out employment forms, W-4 forms, and nondisclosure agreements as well as other documents. Usually, an HR representative presents an overview of the company, the job benefits, and the guidelines and policies that new employees must know.

After this traditional orientation, manufacturing employees should attend some form of product overview training. I call this level 1 training.

New employees need a fundamental understanding of the products that the organization manufactures and markets. Often, this vital training step is skipped, and production workers are simply expected to learn about the products while they are on the job. I believe that hands-on learning is extremely valuable, but I also recommend the addition of formal product overview training. This training would be conducted in a classroom environment, where employees can focus on the information and ask questions.

This training should include real products for employees to touch and analyze. If possible, break each product into its individual components and discuss the part descriptions. Explain how the product is made and why certain parts are assembled onto other parts. Provide a list of the parts and their part numbers, and teach the employees how to read the part numbers and understand what they represent. Typically, part numbers indicate the supplier, the stockroom location, the category of the part (hardware, brass parts, wiring), and so on. If possible, provide a list of all the products that are manufactured. Explain which options are available to customers, and which parts are associated with those options.

Product overview training can be an ongoing process, but it is good practice to provide new employees with up-front product orientation. The length of this training will vary from one company to another. For example, one company for which I consulted created a product overview curriculum that included a one-day initial course and subsequent shorter refresher courses that were conducted once a week for three weeks. Regardless of the content or length, company product overview training is a smart approach to training new employees.

Level 2: Quality Overview

Quality is critical to company success, and therefore it should be emphasized at the time new employees are hired. Level 2 of new-employee training should be devoted to teaching the importance of quality and the fundamental aspects of the company's quality program.

Chapter 3 describes the concept of quality at the source and the importance of placing the responsibility for quality at the point of build. New operators should know that quality is in their hands and that they will perform certain incoming and outgoing quality checks as part of their daily work.

Discuss the top three, or five, quality issues currently being addressed by the organization. Show new employees last year's data in regard to customer and service technician complaints. Allow them to view internal quality information so that they can identify the common factory errors made before product leaves the building. This information should not be a secret. Awareness is a key factor in getting employees to react and act.

Level 3: Introduction to Lean Manufacturing

Lean manufacturing training, from top to bottom in the organization, is a must. Everyone must learn how lean principles are applied in the company and must understand the organization's goals for its lean journey. It is equally important to show new employees that lean is the way of working at your company, without exception, and that the information they are learning is key to this way of working.

Level 3 training is dedicated to teaching new employees the basics of lean manufacturing and, more importantly, how it is being applied in your environment. After this formal training is complete, they will have a better understanding of lean manufacturing and what they should expect when working in a lean environment. Five topics should be covered in this training:

- The seven deadly wastes
- 5S and the visual workplace
- Standard work
- Effective hours
- Kaizen

The Seven Deadly Wastes

The training should provide a description of the seven deadly wastes and the ways they can negatively affect the company. The training must outline and emphasize how working with excessive waste can make day-to-day work life very stressful, especially when workers are expected to meet specific performance metrics in productivity, quality, and volume. Start the training by listing the seven wastes and providing a brief description of each one:

- Overproduction: Building the wrong items, at the wrong time, in the wrong quantity, and in the wrong order

- Overprocessing: Redundant effort or too many checks

- Transportation: Excessive movement of items

- Motion: Excessive reaching, stepping, and walking in and out of the workstation

- Waiting: Time period when manufacturing processes are out of synchronization

- Inventory: Too many parts or partially built products and too many finished goods

- Defects or rejects: Quality errors that require rework and added cost

After providing a good definition of each waste, discuss real-life work examples of each type. Here are examples you can use:

- Example of overproduction: Building more subassemblies than needed. If the process needs 10 wire harnesses, then build ten. Do not build 10 more for the next day even if you have time to do so.

- Example of overprocessing: Sanding, deburring, grinding, and polishing to excess.

- Example of transportation: Moving the product to the next process unnecessarily.

- Example of motion: Leaving the workstation to find parts and tools, or for any other work or nonwork reason.

- Example of waiting: Waiting for the product to arrive in a workstation.

- Example of inventory: Placing a pallet of 5,000 brackets in a workstation when the line uses only 10 brackets per day.

- Example of defect or reject: Reworking a unit when a mistake is made, causing a quality problem.

This approach to teaching the seven wastes is probably sufficient. Regardless of their experience, production workers will quickly understand the concept of waste and will absorb the information. It's a good practice to teach inexperienced production workers the seven wastes, because they will then be able to identify them very quickly, and that is beneficial to the company.

5S and the Visual Workplace

5S training is the most important lean topic for new production workers. In many cases, 5S is the first major lean initiative to be implemented in a company. Once it is implemented in the manufacturing processes on the floor, 5S keeps everything clean and organized, and I guarantee that the existing operators do not want new employees to join the process without understanding 5S. I have witnessed what happens in that case, and it can instantly create animosity toward the new workers. The operators on the line are enjoying the new organization in place and do not want it compromised.

Adequate up-front 5S training will make a world of difference in the development of positive relationships between new and existing production workers. If the company is currently in the process of implementing 5S, the training is still valuable, because newly trained operators can suggest improvements in their new assignments.

Standard Work

Standard work is a bit more difficult to explain to new operators. Typically, they will associate the concept with work instructions or some other form of documentation. As I mentioned earlier, standard work is the best, most efficient, and safest way of performing work.

By itself, this definition may not provide enough insight into what standard work is all about. When you describe standard work, it is important to emphasize that there is a good approach and a bad approach to performing all work. Building a product that requires a lot of movement, walking, searching for parts, and waiting on other processes is not a best practice. When most of the work time is spent actually performing the intended work, it is close to performing standard work. Here is a list you can use in your training:

Standard work is . . .

- Confining work content to the workstation
- Accepting quality responsibilities in the workstation

- Following inspection procedures
- Following testing procedures
- Following safety instructions
- Setting up a machine correctly
- Knowing and following routes and routines for materials handlers
- Following end-of-day cleanup procedures
- Following work instructions

These are all good examples of standard work, and there is a best or recommended way to perform all of them.

Standard work is not . . .

- Leaving the workstation to find parts
- Leaving the workstation to find tools
- Sorting through piles of work instructions
- Waiting on information
- Taking extra breaks
- Talking on cell phones

Standard work is about defining clear roles and responsibilities throughout the factory, with minimal or no waste.

Effective Hours

As I discussed in Chapter 5, effective hours is the amount of time workers spend actually building or assembling product. It is critical that this topic be part of the lean training curriculum. New operators joining the company are not likely to have heard of this concept, so it is important to explain that the lean manufacturing processes are designed to maximize work time.

The training should stress the importance of working together and leaving for breaks and lunch as a team. Explain that workers will be held accountable for specific volume requirements; therefore, using their touch time efficiently is critical to daily success. During this training, give an example of the breakdown of hours, similar to the one outlined in Chapter 4. Remember, you are not operating a labor camp, so be diplomatic in your approach. The key is working smarter, not faster.

Kaizen

The concept of continuous improvement will become a way of life at your company. Lean manufacturing and kaizen are the desired approach to running your business. Therefore, you need to take advantage of the fact that these workers are new and have not yet developed bad work habits. They left those bad habits with their previous employers.

The final component of your lean training is to describe the concept of kaizen and teach workers how they will be involved in kaizen events throughout the year. It is wise to let them know early on that they will be asked to make improvements to the organization. Break down the fundamental aspects of the company kaizen program outlined in Chapter 3.

Teaching the theory of kaizen is challenging. Explain that the kaizen events will be executed often and that workers will be asked to participate. Discuss the purpose of the suggestion box, show them where it is, and explain how to complete the suggestion form properly. Discuss how the suggested ideas will be considered in making improvements to the company. Display pictures of any prior kaizen events, and show them a copy of the kaizen newsletter. Walk the workers around the production floor, identifying the location of the kaizen communication boards and discussing their purpose. Kaizen and lean manufacturing are serious business, so it is important that they be included in initial training.

Level 4: Mock Line Training

In contrast to other kinds of work situations, manufacturing is a dynamic environment. Some new production workers may not have manufacturing backgrounds and may be accustomed to different working styles. For example, retail workers often have times when customer flow is slow. Many workers tend to prepare for the rush, work the rush, and then recover from the rush. In some jobs, the workload levels vary from one day to the next. It's important for workers to realize the distinction between other environments and manufacturing, which is highly aggressive and fast-paced. This is not to downplay the service industry, because it has its own busy times, but manufacturing and service are completely different environments. Service workers bring great knowledge and usually a desirable work ethic. However, they must be adequately prepared for manufacturing work.

Level 4 of new-employee training involves working on the production floor, away from the real manufacturing lines, in what I like to call

mock line training. A mock line is a training area in the factory that looks, and works, like real production. It is important to design and construct this area to be identical to the manufacturing process that workers will be responsible for. This training area should be set up based on the processes in the plant. More importantly, it must contain all the lean programs you have in place.

To design your mock line, begin by identifying an actual assembly process in the plant—something small that can be easily copied and constructed. Create enough workstations to allow the new operators to experience working side by side. Each workstation should contain all the necessary tools to build the product: air tools, parts, bins, fixtures, hardware, work instructions, and safety equipment. The mock line should have all the 5S detail in place: floor tape, station signs, designations, tower lights, and a materials replenishment system. The new operators should be taught how to build real products working within a lean process that is controlled by structure and protocol.

I realize that not all manufacturing processes use assembly-based systems. Therefore, if your company is a job shop or a fabrication-style business, set up your mock line as appropriate for your specific environment. For example, a composite factory, such as a fiberglass plant, should establish a mock line that teaches workers how to do hand lay-up.

New operators can learn the concepts of single piece flow or controlled batches. They can be taught, and then shown, how to flex from one workstation to another. This type of training environment allows them to learn how to work in a lean process and prepares them to go to work when they are assigned to a real production line. Existing operators will appreciate how the mock line ensures that new operators follow the rules of the lean process. It will shorten the learning curve of new operators and will allow them to build a high-quality product for the organization.

Cross-Training Program

At this point, you have some well-trained employees working in your processes. It will take them some time to learn their new environment, but at some point you will want them to learn other stations, operations, and processes. Flexible production operators add significant value to any company, especially in a lean workplace. Operators must be cross-trained, over time, to learn new jobs and operations. This practice allows production supervisors to assess the skill levels of their people and shift workers to different areas of the floor as needed.

All existing employee training curriculums need to have a cross-training component. Cross-training must be customized to fit each company and each process within the company. Here are three considerations for your cross-training program:

- Levels of progression
- Progression of temporary workers
- Cross-training matrix

Levels of Progression

Before creating a cross-training program, you must establish levels of progression, identifying the specific job skills an employee must have in order to move up to another level within the company. Three categories should be established:

- Novice
- Certified
- Trainer

Novice

An operator at the **novice** level is new to the workstation or has just been hired. Essentially, this person is an entry-level employee who has just attended the four levels of training, or an existing employee who has transferred from another area of the company. In either case, the novice employee has passed the initial training process. Depending on the company, new employees may have a 90-day probation period, allowing you to assess how they handle the work. Probation is also a good policy for existing employees who are being trained in a new workstation or area, perhaps with a shorter period of 7 to 10 days.

Novice operators are given some leeway and should be allowed to make minor mistakes while they are learning. Place novice operators with experienced workers so that the novices can be helped if necessary. Determine the amount of time that the experienced workers will assist the novices, and then let the novices work on their own. Monitor the novice's ability to follow the work instructions, perform the required work, adhere to cycle times, and conduct quality checks. Learning curves will vary depending on the complexity of the product, the number of tools, and the abilities of the new operator.

Once a novice has worked to the end of the probation period, you will make an initial decision regarding whether he should be considered for certification status, the next level. Many factors will influence your decision, such as whether he is a new or existing employee as well as his performance during training. Nevertheless, he should remain at the novice level until he has had the opportunity to be a consistent performer for a specified period, which you will decide. The novice must work within takt time (see Chapter 5), perform the work content, and conduct quality checks. He must also react properly to resolve any issues that arise.

The production supervisor should conduct a few audits during this period to see how the new worker is doing. When the worker has proven that he can work consistently, day in and day out, for the given time without supervision, he is ready for certification.

Certified

The second level of the cross-training program is to become certified. **Certified** operators run the show. Although they have the experience and knowledge to run the show and are capable of making good decisions regarding quality errors and flow, they may not be ready to provide training to others.

Working in a lean process takes skill, and that skill is developed over time. Certified operators understand their responsibilities in their own workstation, as well as the procedures and protocol that control those processes. Certified operators may remain at this level for a long time. When are they ready to train others? Naturally, certified operators should strive to reach this next level of job performance on the floor. However, it is wise to first offer them the option of moving to another workstation, or area, before progressing to the training level. Certified operators will become novices in the new area or workstation and will have the opportunity to cross-train in another skill or skill set. Some operators will see the value in that offer and will choose that path.

The cross-training option is wise because some certified operators may not feel that they are ready to train others and may want to continue progressing through the various workstations or areas first. Others will want to move directly toward a training certification status. Of course, management should always persuade the worker to move in the direction that is best for the organization as a whole. But giving operators a choice empowers them to take greater control of their own advancement, and that is very positive.

Both choices are natural progressions for a certified operator. Each choice has its own requirements. If the certified operator decides to try her hand at a new workstation, she will follow the novice level progression guidelines, as mentioned earlier. To become a trainer in a particular workstation, the certified operator must work in that station for a specified period, perhaps three to six months. The exact amount of time depends on your manufacturing processes.

I also recommend that a series of tests be performed on the production line to verify the operator's conformance to standards. Some certified or veteran operators may begin to ignore procedures and lapse into poor work habits; this is human nature and also happens with salaried and support staff. The best method of qualifying certified operators for advancement is to create a testing and auditing system for those who would like to achieve trainer status.

Trainer

Workstation **trainers** have mastered their area. When employees have successfully gone through the new-employee training program, they are placed at a workstation with an operator who has trainer status. The goal of any cross-training program is to get all permanent employees to achieve trainer status.

Training others is almost an art form. It takes time and practice to get comfortable dealing with people and teaching them what you know. Many people really understand a subject or know how to perform a job but cannot relay the information to others in a way that is clear and easily understood. Make sure you know your audience and that the training is delivered in a way they understand it. For those who do not have this capability or desire, the opportunity to cross-train in other lines or departments is a good option.

Progression of Temporary Workers

Temporary workers exist in almost every industry, including manufacturing. Many organizations use temporary workers to help during times of increased output. Certain times of the year are busier than others, depending on products and industry needs. For example, companies that manufacture gas heating stoves are busier from late summer through winter. Construction products such as vinyl windows and siding are manufactured heavily during spring and summer. Rather than employ a larger, permanent workforce year-round, companies bring in temporary workers during the busy seasons.

This approach is also used to identify reliable workers who can eventually become full-time, permanent employees. Because temporary workers often become permanent, a system of progression should exist for them. The goal is to identify the time spent in various activities and the point when they will be eligible for permanent status.

Many companies do not develop such a progression, and temporary workers remain temporary for years. I firmly believe in developing a solid pool of talent, and temporary workers are a wonderful source. Any company embarking on a lean journey should be proactive, identifying early on those workers who have the attitude and ability to assume future responsibilities.

New talent always improves the overall pool. Sometimes, the workers who have been with the company longest and have the most knowledge and experience are the most resistant to change. New workers bring a fresh attitude and approach and have not yet established poor work habits. The blending of these workers is beneficial to your organization, because they can balance each other and inspire and motivate one another. Both types of workers are necessary for successful lean implementation.

Temporary workers are usually seeking permanent employment. They come with new ideas and attitudes and are eager to learn and to prove their worth. Most of them want to become part of your company. It's true that there are also those who simply work for a time and then leave. However, in most cases, temporary workers really want to become permanent, and they will do their best for you. I have seen temporary workers who can outperform many certified operators. I have even witnessed temporary workers training permanent workers.

Cross-Training Matrix

Production supervisors need to be able to identify all their workers, their current responsibilities, their positions in the cross-training program, and their next level of advancement. You should create a cross-training matrix for each manufacturing process on the floor, identifying the workers assigned to the area and the number and type of processes that each worker can perform. The cross-training matrix is a valuable tool for visual management. Figures 6.1 and 6.2 show examples of cross-training matrixes.

2500 LT Line

Operators	Station 1	Station 2	Station 3	Station 4	Station 5	Station 6	Station 7	Station 8
Peter T.	C	N						
Ken L.	T	T	C					
Micheal D.		C	C	C				
Kyle R.			N	N	N			
Lisa P.				N	C	C		
Terri N.					T	T	T	
Mark Z.						N	N	N
Lenny B.							C	N

Levels	
N	Novice
C	Certified
T	Trainer

Figure 6.1 Assembly Cross-Training Matrix

Fabrication Department				
Operators	**Brake Press**	**Spot Weld**	**Stud Press**	**Drill Press**
Lorence R.	C	N		
Rita P.	T	T	C	
Ryan R.		C	C	C
Jenny I.			N	N
Phillip Q.				N

Levels	
N	Novice
C	Certified
T	Trainer

Figure 6.2 Fabrication Cross-Training Matrix

The cross-training matrix provides a useful visual representation of the current skill levels in the area of supervision. This matrix should be posted in the work area and updated regularly to ensure that the information is current, accurate, and can be properly used. Cross-training matrixes for manual assembly are less complex than matrixes for processes that include machines and equipment. Operators working in manufacturing processes with the equipment identified in Figure 6.2 will probably need longer, more structured training.

Training Managers and Engineers

New employees entering into support staff roles also need training that sets correct expectations for work in a lean organization. These individuals should go through the same training curriculum as production workers: product overview, quality training, lean training, and mock line training. Are you surprised? Many Japanese manufacturers require their new managers and engineers to work on the production floor for a certain period of time. It allows them to see how the products they will be supporting are made, and it also encourages the development of working relationships with the operators. Working on production lines also provides insight on issues that create obstacles for operators. Although I am not necessarily recommending this approach, it is certainly worth considering.

Managers

As new managers come into the organization, they should be briefed on the company's lean journey, preferably during the interview. Some companies make lean experience a requirement and list it in the job description for managers. If your company adopts the structure of the company kaizen program described in Chapter 3, your new managers should be trained on this structure and taught how they will be expected to contribute.

The kaizen steering committee is an integral part of the lean program, and new managers may become part of this decision-making group. Employees from their departments will be asked to participate in kaizen events. If the concepts of kaizen and lean are foreign to these managers, they will not be prepared for what takes place. New managers should be aware that their input, contributions, and involvement are required for the continued success of the lean journey.

Engineers

New engineers also play an important role in the lean journey. Manufacturing, industrial, and process engineers will assist with data collection for time studies, waste analysis, process analysis, and value stream mapping. These individuals should go through some form of waste identification and standard work training to learn about the tools used in performing the technical side of lean manufacturing. If the organization has hired a full-time lean or kaizen champion, new engineers should meet this person, because they will help each other capture the current state of individual processes. At some point, the current kaizen champion may leave the process or the company, and new engineers must be prepared to take over this role.

As with job descriptions for new managers, the descriptions for new engineers should reference the company's lean journey, stating its importance. Possibly, new hires will already be familiar with lean concepts and will be perfect candidates to step into their new roles. Although such familiarity is not a requirement, it would certainly shorten the time needed for training. New engineers should also be informed that they will be asked to participate in kaizen events and perhaps even take on a leadership role.

Chapter Wrap-Up

Training is critical. Companies often require salaried staff to attend extensive training programs and to engage in continued education throughout their careers. Their performance plans often cite educational objectives related to salary increases and promotion, and they are expected to keep current on topics in their fields of expertise. Organizations should place just as much emphasis on the training of production workers. Production workers are value-added employees, and an investment in their training equates to profit for the company.

It is difficult for some companies to justify the type of new-employee training I have outlined in this chapter. This approach challenges some of the existing management methodologies and therefore can cause management resistance. However, many companies view training as an important element of the continued success of their lean journey. Upper management must decide whether or not training is considered valuable.

 Often, the issue is simply the lack of a training budget. A good friend of mine operates a small fiberglass manufacturing company. We discussed training and budgets, and I asked him to tell me his constraints. He replied, "I don't need a budget to train my people. If it is needed, we do it."

Well put.

Pia in T sh.

Blue White Nike Air man

Yeah yes going b FAT

I set a taxi to factori

Vant(a Buscar forme'

'Hots the vaic pad'?

the one thes ist seb is it?

I'n   also payn a lot (money & the

Lean Manufacturing as a Growth Creator

As a lean practitioner and consultant, I need to be a salesman, selling the idea of lean manufacturing and kaizen principles to potential customers. This part of the job is not as fun as guiding my clients through their lean journeys. But it isn't difficult to sell the concept of lean; I simply explain how it contributes to growth. Lean is a business strategy, and strategies usually give you ways to improve your company financially and aid in its future growth.

Lean is a companywide approach to continuous improvement, and as time goes on, increasing numbers of employees are involved in the process. Organizations that truly embrace lean manufacturing and continue to fight through the battles of culture change find ways to return the favor to the people in the company who made it happen. Operators, line leads, production supervisors, engineers, and managers represent about 90 percent of the people who do the hands-on work of lean implementation. As the company begins to see the return on its lean investment, there should be a way to trickle some of the profit back to the employees.

Although adding product lines and the processes to manufacture them contributes to growth, giving back to your internal change agents is also critical. In this chapter I outline a variety of approaches for offering lean incentives and the structure that should be in place to encourage this program.

Lean Goals

Implementation of lean manufacturing can have a positive effect on an organization's costs. The rate of return differs from one company to the next, because every approach is unique. Some companies have seen savings of $50,000 to $1 million in the first year of lean manufacturing and as great as $500,000 to $4 million after five years. The smart ones put some of that savings back into the company and in employees' pockets in the form of bonuses.

As a company decision maker, you want to create an exciting energy in your lean program. Part of this excitement is generated when you create goals that everyone will strive to meet. When employees achieve these goals, you should reward them financially.

These goals are the key shop floor metrics outlined in the company's strategic purpose, as described in Chapter 2:

- Productivity

- Quality

- Inventory or WIP

- Floor space use

- Throughput time

As mentioned in Chapter 2, you gain success by improving these metrics. Management sets annual targets, and the employees go after them.

In Chapter 2 I also discuss the importance of finding a balance between cost, quality, and delivery. Some companies use these three indicators as the company metrics. But to operators, they really mean very little. I'm not implying that operators do not understand the concepts, but these three drivers are too far out of their range of responsibility. Even engineers may not know what needs to be done to improve cost, quality, and delivery. My point here is that productivity, quality, inventory, floor space use, and throughput time have much more definable parameters. These five shop floor metrics are directly connected to cost, quality, and delivery, so you can make incentives or bonuses contingent on meeting or exceeding the goals for each metric.

To implement lean principles in manufacturing and assembly processes, you will create kaizen teams and schedule multiple kaizen events. Because of the incentive program, employees will be encouraged to participate in kaizen and try to better themselves and their way of working.

The kaizen teams will be aggressive in their approach. Continuous improvement efforts will accelerate because employees know that there is an incentive at the end. You can use these goals as the catalyst to accelerate your lean journey.

Although lean manufacturing will become your new way of conducting business and employees must be engaged, I would like to see a payback. If an organization books a cost savings of $500,000 in the first year, then it should return some of that to the people who made it happen. It starts with targets and goals, and you must make sure to raise the bar a little higher every year.

Pay-for-Skill Program

As stated in Chapter 6, I am an advocate of employee training. When meeting with clients during strategic planning sessions, I schedule time to talk with the line operators and get their opinions about the journey ahead. The most common topic that is raised during this session is training—specifically, formal training for new employees. I've already outlined new-employee training, but here I want to add an important principle: Any good employee training program must be backed by what is called a pay-for-skill program.

It's true that financial incentives are not the only incentives needed. Chapter 3 shows the importance of recognition and appreciation of employees' lean contributions. The suggestion box, the kaizen newsletter, the communication boards, participation in kaizen events, and kaizen event reports provide incentives on a nonmonetary level.

Beyond that, **pay-for-skill programs** provide financial incentives for workers to learn new jobs in the company. It is in essence a career advancement program for production workers to encourage cross-training on the production floor. For the company, it helps encourage operators to become more flexible and skilled so that the organization can adjust to differing seasonal demands and changes in volume. For the worker, it outlines a clear path for advancement and growth in the organization.

Each process, assembly line, or work cell should have its own pay-for-skill structure. Although some processes may be similar, there may be subtle differences in the jobs and work in each process that warrant a custom program. The cross-training matrix for each process is one of the guides in establishing this type of program, but a lot of detail is involved.

The number of progression levels in a pay-for-skill program will vary and depends on the process. As you develop your levels, I recommend using the following criteria:

- Number of certifications
- Years of experience
- Attendance
- Kaizen and kaizen event participation
- Quality errors

Number of Certifications

Each pay-for-skill level should contain a certain number of jobs that each operator must be certified in. For instance, to fulfill this criterion, an operator might need to be certified in three workstations. Becoming certified in only one workstation does not satisfy the criterion. If an organization follows the rules for new-employee training and cross-training, then the company knows that the operator is experienced. Three workstations is a suggestion; I have seen some companies require certification in five or seven. It's up to the company and the complexity of the manufacturing process.

Years of Experience

The experience criterion should be similar to the timeline for becoming a certified operator after completing the novice portion of the cross-training program. Because there is a time frame during this phase of development (including the number of jobs required in the level), there should be no question about the experience level of the worker. The guidelines for experience are already part of the cross-training program.

Attendance

Absenteeism, tardiness, and turnover problems can create major problems for a company. It is important to develop a criterion for attendance in the pay-for-skill program, not only to encourage people to come to work but also to provide an incentive for doing so. This kind of incentive is sometimes difficult for management to swallow, because they consider good attendance simply an aspect of holding a job. I once worked for a company that had 55 percent turnover among production workers, and that number did not include those who often missed work or showed up late.

Most organizations have defined guidelines for attendance, generally based on a point system. For example, operators might be allowed 10 points every year, and as they miss work, arrive late, or call in sick, points are deducted. A missed day might cost 1 point. Showing up late could warrant a ¼-point deduction, and calling in sick might be worth ½ point. If an employee exhausts all of his points before the end of the year, he is given a verbal warning. The operator is essentially on probation and cannot miss any more days; if he does, he receives a written warning. Any infractions after this point result in termination. Attendance point systems differ from one company to the next.

When you develop the attendance criteria in a pay-for-skill program, you should consider tighter rules. To move to the next pay level, workers should maintain nearly perfect attendance, with some exceptions. Simply staying within the provided points does not warrant promotion. Missing work without advance notice or failing to explain an absence is not acceptable, at least in a pay-for-skill program. Although people can progress by giving advance notice, points are taken from their totals. Another approach is to allow a reduction in points up to 20 percent, notice or not, and still allow a worker to move up.

The point is that you should create a structured guideline for attendance so that the organization moves operators through the pay-for-skill program when it is truly deserved.

Kaizen and Kaizen Event Participation

Kaizen involves everyone in the company, and the organization should encourage active participation in the continuous improvement initiatives. As the company develops its lean culture, employees should be allowed to make improvements to their work areas as often as possible.

In 2006 I worked with a new client, a small family-owned machine shop with gross revenues of about $1 million. The manufacturing floor was about 50,000 square feet, and the company had been implementing 5S for about a year. On my second visit I took a tour with one of the production supervisors. Computer numerical control (CNC) machines, lathes, deburring stations, and inspection areas had been combined into individual work cells. Each cell did a variety of work while staying within a family of products and similar processes.

The production supervisor managed cells 5 and 6. We were walking through cell 5 and watching the action. I had a lot of questions for him, because this cell was scheduled for another kaizen event to help decrease setup times and organize the fixture and jig inventory. He was discussing the flow of the cell when he stopped in his tracks and began to stare down at a desk. This desk was used by the cell's line lead, who gave the machine operators their schedules and day-to-day tasks. (The production supervisors managed the line leads.)

I became curious and asked the supervisor what had caught his eye. He pointed to a small cardboard pen holder taped to the desk. The holder was clearly made from a corrugated box that came from a supplier. This makeshift pen holder contained three highlighter pens: green, yellow, and red. The production supervisor explained to me that the holder had not been there the preceding day, and items like that stood out since the company had embraced 5S. Basically, a new item had appeared on a desk that had everything clearly labeled and identified.

We approached the line lead and asked her why she had made the box. She told us that every morning she received the work orders from the office and it was difficult to see which orders had to be worked on first. Some jobs could be done much quicker than others, and some had longer delivery dates. The line leads were empowered to distribute work as needed to ensure that all orders were complete on time while minimizing wait times and imbalances between machines.

After she received the work orders, the line lead spent about 30 minutes each morning sorting through them before handing them out to the line workers. The line workers then prepared their work areas as needed.

The purpose of the three highlighters? She used them to color-code the work orders to quickly show their order of importance: Green indicated rush or fast jobs, yellow identified orders with longer completion dates, and red indicated the least urgent work orders of the day.

By making improvements to her area, the line lead was practicing kaizen. It is this type of mentality that is needed. In regard to the pay-for-skill progression, a company could document this kind of small improvement throughout the plant and keep records of those operators or other floor personnel making the changes. Each level in the program might require a certain number of small implementations as part of the required progression into a higher-paying level.

The company kaizen program encourages workers to offer improvement ideas by using the suggestion form. Advancement to another level could simply require production workers to submit a certain number of suggestions.

Probably the most important part of the kaizen and kaizen event participation portion of the program is the number of kaizen events in which the operator has participated. This part of the pay-for-skill program is somewhat difficult, because scheduling the events and selecting the teams may differ from one year to the next. If an organization conducts only five kaizen events in a given year, the chances are that some production workers may not get the opportunity to be selected. Ideally a company should strive to have at least one kaizen event each month; the reality is that the number of events could be lower. It depends on how aggressively you are pursuing the lean journey. When you have achieved consistency in scheduling kaizen events, then you can make it a requirement for progression to the next level in the program.

Quality Errors

Production workers are responsible for quality and for the implementation of quality at the source. An organization can easily track line errors, but it is wise to remember that operators are human. They are faced with challenges every day that can make it difficult for them to do a perfect job. However, if production workers repeatedly make mistakes, the process needs to be analyzed further to reduce the occurrence.

I am a firm believer that the company must effectively design and set up its processes so that the production workers have all tools needed to be successful. Once the process is in their control, then errors should be at a minimum. If a production worker continues to make mistakes, it should be tracked and reviewed. One of the requirements for advancement in regard to quality might be the number of mistakes made in the process. The goal here is not to point fingers at people but to encourage strong performance and reward workers for meeting certain performance standards.

Figure 7.1 shows an example of a pay-for-skill program. It outlines the requirements for workers to advance to each level. As you can see, the

Pay for Skill			
Level 1	**Level 2**	**Level 3**	**Level 4**
Certified in 3 Stations	Certified in 5 Stations	Certified in 7 Stations and Trainer in 1 Workstation	Certified in All Stations and Trainer in 3 Workstations
90 Days Experience	90 Days Experience	90 Days Experience	90 Days Experience
2 Days Missed in 90 Days	2 Days Missed in 90 Days	1 Day Missed in 90 Days	0 Days Missed in 90 Days
1 Kaizen Event	1 Kaizen Event	2 Kaizen Events	3 Kaizen Events
3 Quality Errors in 90 Days	2 Quality Errors in 90 Days	1 Quality Error in 90 Days	0 Quality Errors in 90 Days

Figure 7.1 Sample Pay-for-Skill Requirements

Pay Progression			
Level 1	**Level 2**	**Level 3**	**Level 4**
$12.50	$13.45	$14.25	$15.75

Figure 7.2 Sample Pay-for-Skill Pay Opportunities

requirements become stricter and require more work and dedication from the operator as advancement occurs.

Figure 7.2 shows the pay increase opportunities that are available when an operator completes the requirements.

A pay-for-skill program can be very successful for a company. It can also be a headache and can create some animosity between workers. The point of this program is to provide a clear career path for production workers, who are highly valuable employees because they build the products that financially support the organization. Companies usually provide career advancement opportunities for managers and engineers, but little effort is given to providing one for production workers. Consistent performers and lean change agents should be rewarded, and a customized pay-for-skill program is a good approach.

Providing Incentives for Good Ideas

Some companies that I have helped have taken their incentive programs to a higher level. Although acknowledgment and praise are best in the long term, you can add yet another monetary incentive program. As a lean company, you want to continue to encourage production workers to come up with continuous improvement ideas all the time. This is kaizen. As I've mentioned, the suggestion form is a great way to garner fresh ideas from the production floor, but there is another way to get workers to make changes.

These improvements may take some time, because they are implemented as the days and weeks progress. If an operator sees an opportunity to make an improvement to the process, implements the improvement, and it reaps a financial savings or gain for the organization, you should award her with a check of some value.

For instance, suppose that a production worker sees a better way to package the products to reduce time and material. The recommendation should be reviewed by management and engineers to see whether it is feasible from a product specification perspective. After approval, the idea goes through an engineering change request process, and then the new process change is initiated. After a given period of time, the improvement can be measured to see whether it has made a positive impact on productivity and material cost. You can amortize the savings over a year to calculate its annual cost savings. If the improvement saved the company, say, $20,000 annually, the employee might receive a check for $500.

In the beginning, this kind of program will be slow to encourage workers, but after the first or second idea turns into a paycheck, more workers will be excited and will begin their own improvement projects. These projects, of course, involve other employees, but idea generation is the start. In addition, with a program in place to encourage the behavior, more production workers will become engaged in the process. Each company must establish guidelines for the program in regard to annual savings, time frames for implementation, and the amount of money to be paid to the worker. It is another approach to soliciting continuous improvement ideas from factory floor workers.

Chapter Wrap-Up

Financial incentives in a lean journey are not the sole approach to thanking people for becoming multiskilled and contributing to continuous improvement efforts. Because lean manufacturing is a business approach, it essentially is a job requirement for working in a lean organization. But one of the reasons lean implementations fail on the shop floor is that the company does not recognize the contributions that can come from production workers. Many organizations are taking major steps to reward their best production workers and provide incentives for them to assist in implementing a lean program. Because lean manufacturing is truly a growth creator, you have numerous options to reward those who help you in your lean endeavors.

Lean Leadership
Made Simple

Assuming the task of leadership can change you positively or negatively. It can make you a demon or a saint, and it will make or break your effort to implement lean manufacturing in your organization.

A colleague of mine used to be a plant manager, and he earned a reputation for treating the people who worked for him very well. After several yeazars as their manager, he was promoted to a vice presidential position with the same company. Six months into his new position, we got together for lunch to discuss my plans to write this book. My friend was a much different person from the man I remembered. Corporate life had taken control of his identity, and he seemed genuinely disappointed with himself.

I remember clearly one of the things he said to me that day: "When you accept an executive position in corporate America, you have to leave all your ethical reasoning behind you." He explained that while working at the plant level, he still had some control and was able to create a pleasant work environment for his people. Now, as an executive, he had to play the role of a greedy, selfish, ruthless businessman. Although he realized that he could simply walk away from the company, it wasn't easy for him to do because the money and career opportunities were very beneficial. Therefore, he found himself in a continuous struggle, trying to maintain his former value system in a position that was in total opposition to his personal beliefs.

I realize his situation is not representative of all companies, but it got me to thinking about guiding employees through a lean journey. Can a company's past behavior toward its people affect the success of a lean effort? Are employees forced to work excessive overtime and placed into positions that make them unsuccessful and over time create a negative culture? My friend's company was beginning its lean journey, and he was experiencing a conflict between the way the leaders operated the organization and the way they were being taught lean manufacturing by lean practitioners like me. Lean journeys require investment, time, commitment, patience, and tolerance of mistakes. Managers controlling the day-to-day operations of the company were far from accepting this approach to business, and my friend eventually found himself looking for another job. It was a personal choice.

My experiences over the past ten years in the lean field have taught me a lot of valuable things, especially about how to treat people. The companies I have assisted quickly realized that a new approach to leadership would be needed if they were to ensure success in their lean endeavors. I was, by no means, a perfect employee in the years before I founded my consulting company, Kaizen Assembly. In fact, I was a bit resistant to lean. However, I always believed that my resistance was normal, and I appreciated my great lean leaders. The most important principle I've learned is that the way we treat people in our lean journeys is the cornerstone of lean leadership.

I now use what I've learned to lead companies in a manner that seems fair and just, and I hope that this approach trickles down through their organizations. Organizations embarking on a lean journey need effective leaders who understand the importance of employee contributions and realize how much their efforts and attitudes affect the company's success or failure. Certain corporate leaders need to realize that even though aggressive practices may achieve short-term financial success, they also place the company on the path to a precarious future.

My colleague's perspective on leadership was altered dramatically after only a few months in an executive position. Although he realized the negative personal changes that were occurring, he simply had no choice except to acquiesce—to conform. But when lean was brought into the picture, he knew it was time to leave.

Many executive leaders are breeding a middle management culture that is willing to sacrifice the rights of employees. This is an opinion developed from observations I have made. Again, not all leaders, but a good handful of them lead their organizations with too much negative reinforcement. Profits are necessary, but to help deal with the culture change in a lean organization, a mind-set that places profit before improvement can become an obstacle to success. Once the management level of a company has been indoctrinated, younger leaders are then trained to be loyal to the company at all costs, relinquishing their personal lives. The company becomes their lifeblood, and their identity becomes defined by a prestigious title and by their "loyalty and dedication" to the company (which translates into how much of their personal lives are sacrificed).

Poor Leadership Traits

Lean leaders are only human beings; therefore, they typically conduct themselves in a manner that reflects their authentic personality. If an individual is generally grumpy and negative about change, his management techniques will demonstrate that, and he will affect the morale of others through his body language as well as his words.

In contrast, individuals who are happy and positive tend to lead in a happy and positive manner. Lean leaders who do not let negativity influence their actions will create a following of positive thinkers.

Because leaders' personalities are reflected in their management techniques, poor leaders can be categorized as follows.

The Master Delegator

This is a manager who accepts the credit for work done by her group but doesn't accept the blame when things don't go as planned. I experienced this management style when I worked for someone who took credit for my successes by informing superiors that I had done the tasks under his direction. On the other hand, he was sure to point out any mistakes that I made, especially when we were in the company of his fellow managers.

A delegator is a master at setting up others to fail, because it diverts attention from her and disguises her own inabilities or poor work ethic. Delegators often give the impression of overwhelming self-confidence, but in reality they usually have very low self-esteem and a fear of failure.

People who answer to this type of manager should exercise caution, because the moment they show any sign of being able to multitask, they will quickly be assigned many projects and possibly overwhelmed—the more they do, the fewer responsibilities for their boss.

The Yes/No Manager

A yes/no manager appears to be busy all the time and yet rarely accomplishes much. This type of manager doesn't care to engage in meaningful, intelligent discussions because it robs time from his schedule and requires him to think. Therefore, he reduces all output to a simple yes or no answer, something that can be very annoying to employees who are seeking guidance. A boss is responsible for leading his employees and providing assistance when required. Simply answering yes or no indicates to employees that the boss has no time for them and that they are not valued. Often, this manager's "busy bee" buzzing is a smokescreen to hide the fact that he rarely does anything.

The Crisis Junkie

Subordinates must be on their toes with this one, because under this type of manager they will work late at night or on the weekends (or both). A crisis junkie typically postpones acting until the last minute and then panics and starts hunting for someone to help her meet her deadline. An individual like this lacks time management skills as well as a true understanding of the job or department she is responsible for. Workers are often required to drop what they are doing, regardless of their own deadlines, to assist her in meeting hers. Everything is of equal—and utmost—importance, and her panic indicates that. Of course her subordinates want to help her, but they should be warned that often little planning has been done, and the project is a disorganized mess. It would be funny if it weren't eating up so much of the workers' personal time.

The Poor Decision Maker

These managers seem to have obtained their positions of power by default. There are many reasons they can never seem to make a decision completely on their own. One reason is that they are probably incompetent. Another reason, and probably the most common, is that they fear making a bad decision. They take input from everyone around them and do not make a move until they have polled all possible sources. These poor decision makers can also be mistaken for crisis junkies, because they often wait until the last minute to place a call for help, and then subordinates are in the position of following them around while they hover and panic until the job is completed.

The Personal Boss

This type of manager cannot seem to separate his personal life from his work life, and more often than not, his life is dysfunctional outside the work environment. Everyone has life issues that can be upsetting, but a professional makes an effort not to bring them to work.

A More Congenial Leadership

In a way, many leaders have lost a sense of reality. After all, treating people ethically is not like performing open heart surgery. You don't need week-long seminars and workshops to learn how to be nice.

Although leading a company is a monumental task, leading a group of employees is relatively easy. I enjoy leading people because of its simplicity. What follows is my concept of how to make lean leadership easy and pleasant for yourself as well as for those you lead.

Acknowledge and Involve Your Staff

When a member of your staff does a good job, notice it, and praise her for a job well done. Many managers don't praise employees, because they feel employees are paid to do their jobs well and praise is unnecessary. That is a cop-out. It doesn't require much effort to say, "Good job," and that is all that is required. Don't miss an opportunity to praise a piece of good work.

Go to your employees for advice, and engage them in problem-solving issues. Using the talent that is available to you is a key ingredient in building a positive and helpful team. One person may have the solution, or perhaps members of the department can resolve the problem together. In either case, involving your workers promotes trust as well as professional interaction.

Provide an Environment in Which People Can Be Successful

Train your employees adequately, and give them all the tools they need to be successful. Explain all job responsibilities clearly, and encourage questions and feedback. New employees need your attention and deserve your support and encouragement while they are learning. Don't leave them hanging or looking for answers, because it will indicate that you are not a manager they can rely on. Be sure to spend sufficient time getting to know each individual contributor so that you are able to assess everyone's skills accurately. In this way, you can make sure that people are assigned appropriate responsibilities and are challenged and inspired by their work.

Do Not Humiliate Anyone Who Works for You

If you are annoyed with someone on your team or if he has done something wrong, keep your cool and bring it up when you are alone with him. Embarrassing your people in front of others does not show that you are a good leader, but instead exposes you publicly as a tyrant. Always wait until you have an opportunity to discuss the sensitive issue with the employee in private.

Create an Environment Where Mistakes Are OK

Mistakes are learning experiences and should be embraced as such by leaders of an organization. Typically, mistakes are made when an employee is in a learning curve. Beware of those employees who never make mistakes, because it usually indicates that they are not stretching or reaching out for new opportunities. Treating mistakes as a part of growth, rather than something to be ashamed of, allows your employees to feel comfortable taking risks on your behalf.

Remember Personal Details

Spend adequate time getting to know your employees. They have a variety of interests outside work that if known may open the door for better relationships between staff and management. Showing interest in your team as people, and not just as workers, sets you apart as an effective and well-liked leader. Who knows? You may even find a new golf buddy in the process.

Don't Hide behind Your Position

Be genuinely friendly with your people. Put aside your concerns and issues to simply say, "Good morning. How was your weekend?" A morning greeting may not seem like much, but it makes a world of difference in the work environment. Don't hide in your office during the day, but instead make yourself visible and available. Stop by the cubicles or offices informally to say hello or find out how everything is going for each of your people. You would be surprised by what can be gained from this type of interaction. Standing together as a team allows people to support and encourage each other when times are tough.

Be Approachable

Maintain an open-door policy. Allow your staff to come to you whenever they need to talk about sensitive issues, difficulties outside work, or even simple smalltalk. This is one of the most valuable leadership tools I learned from my manager at my first manufacturing job. Whenever I walked into his office with questions or concerns, he would stop whatever he had been doing and give me his full attention.

Admit Your Mistakes

If you are wrong, admit it. Managers are not perfect—they are human—and showing human qualities and frailties is a plus in any organization. Good leaders always take responsibility for their errors and never blame personal mistakes on others on their team. Demonstrating that errors are an expected part of the experience allows employees to feel less threatened and to respect you as a leader. Your team learns that honestly admitting mistakes is the best way to strategize and identify effective remedies or solutions.

Listen in a Way That Encourages Employees to Talk to You

Management intimidates many employees, so good listening skills are crucial to promote honesty and open communication among your team. Make sure you listen whenever an employee needs to share, and show her that you are willing to listen by stopping what you are doing and giving her your full attention. Don't prepare your answer while she is talking. Instead, let her finish her dialog, and then think about what she has said. If you need time to provide feedback, ask her whether you can get back to her. If you do have a ready answer, you can tell her after she finishes talking. Let people know that they are important and worthy of your time, and don't be too busy to listen.

Be Clear in Your Requests

It is your responsibility to communicate effectively to your team members so that they will be certain what you need from them. Clear direction will enable them to get the job done efficiently and with fewer interruptions or confusion. After delivering a message, always ask whether your team members have understood everything or whether they need further explanation or clarification. No one likes to be given poor direction or misleading information, because it makes the job infinitely more difficult and increases the chance for error. Remember, your job is to facilitate work, so communicate clearly.

Stand behind Your People

Supporting your team is a critical part of good leadership, and it can be challenging, especially when an employee fails. Nevertheless, it is important that your team feels it has your support in all circumstances. If a worker feels that you will not stand up for him, then you have failed as a leader. One manager I knew was not seen as an effective leader because he sacrificed his team whenever something went wrong, even if it meant he needed to lie. If one of the team made a mistake, he left the bumbler standing alone in the cold. This is exactly the opposite of how a good leader should handle this situation.

Be a Good Communicator

Employees respect a manager who can articulate what they did wrong without blaming them. Often, mistakes are made because of unclear direction, so look to yourself first for possible reasons for any mishap. Admit your mistake and responsibility first before explaining theirs.

Employees look for a leader who is not secretive and who will pass on important information about the company. Honest communication blunts the power of the rumor mill, and it promotes trust among your team.

Effective Lean Leadership

I have been surrounded by all types of leaders my entire life. I have participated in team sports since childhood, and that gave me many opportunities to lead and be led, as did my many years in academic institutions. I have always had a healthy respect for effective leadership, and I know that people like to be led, to have structure and discipline, and to be asked to do things that have importance and relevance.

Poor leadership of your lean program results in lack of motivation, poor performance, high absenteeism, and, ultimately, high employee turnover. Poor leaders are easily recognizable because they have all or some of the following characteristics:

- They practice negative reinforcement.
- They are focused on their own personal needs rather than the professional needs of their team members.
- They are pessimistic rather than positive.
- They are poor listeners.
- They lack motivation.
- They are closed to new ideas.
- They are slow to adapt to change.
- They blame others rather than take responsibility.
- They provide unclear or uncertain direction.
- They have no idea who their people are.
- They are secretive.
- They are seldom available.
- Their door is usually closed.
- They fear failure.
- They do not stand behind their people.
- They have difficulty developing their employees.
- They exercise leadership by control, manipulation, and coercion.

In contrast, effective lean leadership is not based on control, coercion, and manipulation. Lean leaders are focused on the future rather than the past. Lean leaders gain respect by their ability to inspire others to work toward specific goals. Effective leaders help others to become better people and create workplaces that attract good individuals and keep them happy and motivated to excel.

The first step in being a successful manager is to admit that you don't have all the answers. Admitting that you are not all-knowing gains the respect of your employees, as well as their trust. Being realistic is also a positive characteristic. Realize that it isn't possible for you or your team to solve every problem that exists, and know how to establish boundaries for yourself and your organization. Be yourself, be authentic at all times, and remember that any failures will be forgiven if you are honest and always try to do the right thing.

Many leaders lack fundamental leadership skills. I believe that leaders are born, and managers are trained. Businesses have made gross mistakes by trying to turn managers into leaders, something that is not always possible because leadership characteristics cannot always be learned. The major distinction between leaders and managers is that leaders recognize the value of people, whereas managers drive business. Managers are listened to, but leaders are followed.

Poor leaders create a variety of problems for a company. Often, their actions require operators to work overtime in a work environment that is already unpleasant, because they simply don't understand the problem and don't know how to solve it effectively. Many poor leaders cannot identify skill sets in their organization, and therefore work is assigned to the wrong persons, resulting in hampered morale, poor production, and eventually lost revenues. Lost revenues lead to downsizing and layoffs, which then prompt poor leaders to burden the remaining staff with unreasonable workloads and more responsibilities that don't match their skills, education, or training.

Assigning work to employees who cannot perform it creates overtime, as well as confusion and anxiety for those trying to achieve the company goals in unrealistic situations. Work should be distributed based on the skill level of employees, and leaders should develop plans to provide the appropriate tools and resources to those on the team who need more development. This is not an exact science and not an easy task, especially when companies are focused on the bottom line. However, good leaders are attuned to their people and know how to keep workers doing the right thing—for themselves as well as for the company. Effective leaders know that every employee has something to offer and acknowledge their responsibility to use that talent with projects and assignments that correlate to their ability.

Ten Signs of Incompetent Lean Leaders

How do you know when you are dealing with flawed leaders? Often, poor managers can leave the impression that they are valuable. Human resource professionals have the difficult task of identifying poor leaders within a company. It can be a painful activity, ending in no results.

I have developed a checklist that can help pinpoint those who may struggle to become lean leaders. I call it "Ten Signs of Incompetent Leaders."

Incompetent leaders do these things:

1. **They delegate work rather than balance workloads.**

   This practice diverts attention from the leader in case of failure. The leader may feel that he is managing his people, but in reality he is creating work imbalances within the group. These imbalances can create unnecessary overtime for some workers, and underuse of others. A good manager is aware of his people's skills and allocates work accordingly, to exploit their talents as well as promote learning and career growth.

2. **They reduce all answers to yes or no rather than explain their reasoning.**

   This is an example of a crisis manager who cannot think farther ahead than a few hours. A yes/no manager finds it a waste of time to discover the real answer through intellectual effort. She is already thinking about the next crisis.

3. **They do not separate their personal lives from their professional lives.**

   This manager brings his personal problems to work. Working for this type of manager can be dramatic. He does not set aside his emotional troubles while trying to manage people. He is less focused and does not give you the attention and direction you need for success.

4. **They are always managing crises.**

   If you are a company that has crisis managers, then you can say goodbye to innovation and progression. Proactive thinking is critical to the success of any company. If you are not finding ways to stop or reduce the amount of crisis that must be managed, then your competition will pass you by.

5. They create an environment where mistakes are unacceptable.

Being held accountable for wrong decisions is a fear for this kind of manager. I use the analogy of a basketball player who has no fouls. If he is not going for the ball and taking chances with the opponent—actions that risk being called for a foul—then he is not trying hard enough. Managers need to take a chance and not be afraid to make a mistake.

6. They humiliate or reprimand an employee in front of a group.

This is a clear and visible sign of a poor leader. A good leader takes employee problems away from a group setting to a more private one. If you have a boss who berates her people in public, it is time for a visit to human resources.

7. They do not stand behind subordinates when they fail.

Never leave your people out to dry. Always back them up—right, wrong, or indifferent. If an employee tries his best and fails to come through, he should be commended on his effort and not punished for the failure.

8. They encourage hard workers, not smart workers.

I am not impressed with hard workers. A hard worker is usually defined by the number of hours she puts in. Smart workers are the ones I hire and embrace. Smart workers understand the concept of time management and multitasking. Poor leaders miss this connection. Smart workers are methodical in their thinking and can generally be successful because of their abilities to manage many projects at a time. Hard workers may take twice as long to do the work. It is important to assign work according to people's skills and personalities to ensure their success.

9. They judge people on hours, not performance.

This is similar to number 8. Again, I am not impressed with overtime junkies. They have lost all perspective on a healthy family–life balance. Bad managers promote employees who work the most hours and discard the smart ones who work less, meaning that they have better time management skills. Stop watching the clock.

10. They act differently in front of their leaders.

This is an indication of low self-confidence. Such managers have doubts about their own ability to lead, and they act like little children when authority is present. A confident person acts the same around everyone. Effective managers have respect for their bosses but also have self-respect.

Five Lean Leadership Rules for Success

The challenges faced by leaders are tremendous, especially when an organization begins its lean journey. I have always felt that people are crucial to the success of a company. My lean leadership approach is different from traditional management techniques, and it can be defined in five simple rules. I call it "Five Lean Leadership Rules for Successful Lean Implementations."

1. Hire people who have passions outside work.

It is good to surround yourself with employees who understand the importance of family and leisure life. Generally, people who have passions in life have a good work ethic. I want well-rounded employees on my staff, because creating an environment with diverse people is exciting.

2. Do not hire workaholics.

I am not interested in hiring workaholics. Workaholics create imbalances within a group. My people are allowed to have a life outside work, and their colleagues should respect those lives. If you have imbalances, you create animosity between people. Tension can build because those who work excessive overtime will start to question the loyalty of others. I do not want to hear someone say, "Leaving already?" I do not encourage this type of behavior. In addition, I work very little overtime myself. I do not want to give a bad example to my people.

3. Create a comfortable interview.

My interviewing process is centered on job candidates as people. I ask prospective employees what kinds of interests they have outside work. Do they ski? Are they affiliated with organizations and clubs? I ask them what kinds of hobbies they enjoy.

I try to create an interview process that encourages a sense of self-worth and that recognizes the importance of their personal lives. Toward the middle of the interview, I talk about my family and how I enjoy being home with them. The positions in my department are highly technical, and job candidates come to me with all the necessary skills needed to perform the work. My job should be to balance their workloads to ensure that they can enjoy life away from work.

4. Be a result-driven leader.

When I was corporate lean champion for a large manufacturing company, I tried to be a result-driven leader, and I did not care about hours. Most of my employees were salaried and I paid them for results, not time. If someone could get her work accomplished in 35 hours a week, great! If she felt that she needed to work 45 hours one week to get caught up on an assignment, so be it. However, it was my responsibility as a leader to identity my employees' positive qualities and to balance their workloads so that they could be successful and get away from work to enjoy their lives.

5. Create an efficient workplace.

It is difficult to have a balanced work environment when the company operates inefficiently. Create structure and organization within your department so that people know precisely what needs to be done. My employees did not walk around looking for things to do. Their assignments were given to them every week. An inefficient workplace automatically creates overtime because employees participate in wasted work.

Following this philosophy allowed my department to have the lowest turnover and absenteeism in the company. My people were at work and on time every day. They were able to leave work for family emergencies and could adjust their hours as needed as long as their work was complete. I was result driven, so I expected my people to complete their work as required, and if they slipped it was addressed. They were given a lot of freedom, and when that freedom was abused, it was dealt with in a positive, professional manner. Turnover is expensive, and I did not have the time to constantly be hiring people.

I encourage you to try this lean leadership philosophy. You will be surprised by what can happen.

Poor leadership surrounds us. It is a fact of life. It is unfortunate when employees stop caring because of bad managers. I implore you to challenge your managers to become better at what they do. We as leaders need to realize that people are the number 1 asset in a company. Stockholders do not show up to the company to work; the employees do.

Being a lean leader requires very good balancing skills, because we have a tremendous amount of responsibility not only to improve the bottom line but also to be ethical in the manner that we improve it. Through ethical and sane leadership, you can help your company grow.

Chapter Wrap-Up

I felt it was fitting to place this chapter at the end of the book. The book provides a number of guidelines, so it is important to talk about leadership in general. Often, lean practitioners see issues in management very early during their partnerships with other companies. We recognize that employees are already overloaded with work or report to a poor leader. They can already see problems or resistance from their superiors. How you approach your people and help them embrace the change to lean manufacturing starts with identifying how you treat them.

Lean leadership is nearly impossible to teach, but I hope I have shed some light on the obvious. I want your journeys to be successful. Lead as needed.

Quick Reference

The Seven Deadly Wastes

In Chapter 3, I discuss the importance of communication in a lean organization. Lean manufacturing is a journey, so there must be ongoing information and training on lean principles and the importance of lean manufacturing to the company. To enhance communication about the program, you can use the information in this appendix to create your own poster and display it throughout the company.

- Overproduction: Building the wrong items, at the wrong time, in the wrong quantity, and in the wrong order

- Overprocessing: Redundant effort or too many checks

- Transportation: Excessive movement of items

- Motion: Excessive reaching, stepping, and walking in and out of the workstation

- Waiting: Time period when manufacturing processes are out of synchronization

- Inventory: Too many parts or partially built products and too many finished goods

- Defects or rejects: Quality errors that require rework and added cost

Key Elements of a Company Kaizen Program

A company kaizen program is an essential policy for lean implementation. Following are the key elements of a kaizen continuous improvement program:

- Kaizen champion
- Kaizen events
- Kaizen steering committee
- Kaizen event tracking and scheduling
- Kaizen event communication
- Monthly kaizen meeting

appendix

Supplemental Material

5S Audit Form

Chapter 5 explains the importance of the sustaining portion of the 5S program. Figure B.1 is an example of a 5S audit form that you can use to monitor the shop floor's compliance with the 5S program. It is a template that should be tailored to your company's needs. Simply list the key criteria in each S that you wish to check for in your weekly audits.

5S Tracking Sheet

Similar to the 5S audit form, Figure B.2 is another template that should be customized based on your manufacturing processes. It will help track the weekly 5S scores from the audit and allow everyone to see how each area is conforming to 5S. Make sure to post this form throughout the factory.

Time Study Sheet

The form in Figure B.3 is used to capture time and motion study information in a given manufacturing or assembly process. It will help you identify waste-removal opportunities and allow you to distribute work content evenly throughout the process for better flow. Chapter 5 outlines the fundamental rules in time study collection and line balancing.

5S Audit Form

Team			
Audit Date	**# of Yesses**	**/15 =**	**%**
Auditors			

Sort (Get rid of unnecessary items)

Workstation and/or area is clear of all non-production required material.	Yes	No
Unnecessary equipment and machines have been removed from the area.	Yes	No

Straighten (Organize)

Items on the floor are clearly marked with floor tape and labels.	Yes	No
All equipment and tools are clearly marked and well organized.	Yes	No
Locations and containers for items, parts, and supplies are clearly marked.	Yes	No

Scrub (Clean and solve)

Floors, work surfaces, equipment, and storage areas are clean.	Yes	No
Garbage and recyclables are collected and disposed of properly.	Yes	No
Excess pallet and packaging materials are cleared out of area.	Yes	No
Machines and equipment are free of grease and dust.	Yes	No

Standardize (Tasks)

Standard work is displayed.	Yes	No
It is obvious through visual management whether tasks have been done.	Yes	No

Sustain (Keep it up)

Standard work is being followed.	Yes	No
Work instructions are displayed with correct revision.	Yes	No
Work area is clean, neat, and orderly with no serious unsafe conditions observed.	Yes	No
End-of-day cleanup procedures are posted.	Yes	No

GREEN = 81% to 100%
Area is 5S compliant

YELLOW = 66% to 80%
Area meets minimal standards

RED = 0% to 65%
Area needs immediate attention

Figure B.1 5S Audit Form

AREA	W.E. 9/10/2004	W.E. 9/17/2004	W.E. 9/24/2004	W.E. 9/31/2004	W.E. 10/7/2004	W.E. 10/14/2004	W.E. 10/21/2004	W.E. 10/28/2004
Cell 1								
Cell 2								
Line A								
Line B								
Office								
Receiving								
Materials								
Shipping								
R&D Lab								
Engineering								
Maintenance								
Paint Room								

Green:
81%–100%
AREA IS 5S COMPLIANT

Yellow:
66%–80%
AREA MEETS MINIMAL STANDARDS

Red:
0%–65%
AREA NEEDS IMMEDIATE ATTENTION

Figure B.2 Tracking Sheet

Station/Line/Process

Time Samples

Step	Task Description	VA	NVA	1	2	3	4	5	6	7	8	AVG

Figure B.3 Time Study Sheet

Glossary

5S

A methodology for organizing, cleaning, developing, and sustaining a productive work environment.

- Sort: Remove all unnecessary items from the work area.

- Straighten: Organize what is needed so that it is easily identifiable in a designated place.

- Scrub: Clean everything.

- Standardize: Follow consistent best practices.

- Sustain: Maintain the improvements, and provide opportunities for additional improvements.

5S audit form

A form that is used to monitor the 5S program and to ensure that improvements are made when there are deviations from the program. Audits should be performed weekly to ensure that the program is sustained.

5S tracking sheet

A visual tool that is displayed on the production floor to show how each process or area is sustaining 5S. The scores come from the weekly 5S audits, and the tracking sheet is updated once a month. Incentives should be provided to the process or area with the highest monthly score.

controlled batches

A method of inventory and volume control to ensure that the right amount of product and parts is being built when needed. Operators build to the specified batch quantity; then they stop and verify the quantity and quality.

cross-training matrix

A management tool used to monitor operators' skill levels within an assembly line or other manufacturing process. There are three skill levels: novice, certified, and trainer.

effective hours

The amount of time production workers actually spend building products and fabricating parts. It excludes meetings, breaks, lunches, end-of-day cleanup, and any other scheduled time away from the manufacturing process.

kaizen

Japanese word for continuous improvement. Kaizen encompasses the ideas of encouraging employee participation and promoting a process-oriented culture.

kaizen champion

An employee who is 100 percent dedicated to implementing kaizen and driving the continuous improvement efforts within an organization.

kaizen event

A planned, scheduled process improvement project intended to implement lean manufacturing principles. Kaizen events are planned four weeks in advance to ensure 100 percent participation of team members and achievement of the event goals.

kaizen steering committee

A group of managers and operator representatives that oversees all kaizen event activities in a company. The committee is led by the kaizen champion and meets once a month.

kaizen suggestion box

Used for collecting employee recommendations on continuous improvement ideas.

kaizen tracking sheet

A spreadsheet that is used by the kaizen steering committee to plan and track all kaizen events in the organization to ensure completion of all kaizen activities. See *kaizen steering committee.*

pay-for-skill program

An incentive-based program to encourage multiskilled workers. As production workers learn new jobs, they receive a pay increase for becoming proficient in more areas of the company.

seven deadly wastes

- Overproduction: Building the wrong items, at the wrong time, in the wrong quantity, and in the wrong order

- Overprocessing: Redundant effort or too many checks

- Transportation: Excessive movement of items

- Motion: Excessive reaching, stepping, and walking in and out of the workstation

- Waiting: Time period when manufacturing processes are out of synchronization

- Inventory: Too many parts or partially built products and too many finished goods

- Defects or rejects: Quality errors that require rework and added cost

single piece flow

The movement of parts or units in manufacturing processes one piece at a time.

standard work

An agreed-upon set of work procedures that establishes the best and most reliable methods and steps for each process and each employee. These methods are clearly defined, represent best practice, and are supported by documentation.

strategic purpose

A lean manufacturing business strategy used as a guideline for implementations and improvements. The strategic purpose is revised once a year.

takt time

German word for "rhythm." The time in which a unit must move from one workstation to the next to meet the required daily output. It represents the product completion interval for a given process.

three main drivers

Cost, quality, and delivery. Companies must operate under conditions that offer a competitive balance between these three main drivers. It is difficult to accomplish, because each customer has different needs in regard to these drivers. Application of lean manufacturing principles can help an organization get close to an optimal balance.

tower lights

A color-based light system that is installed at a workstation to allow operators to communicate with support staff and materials handlers.

Index

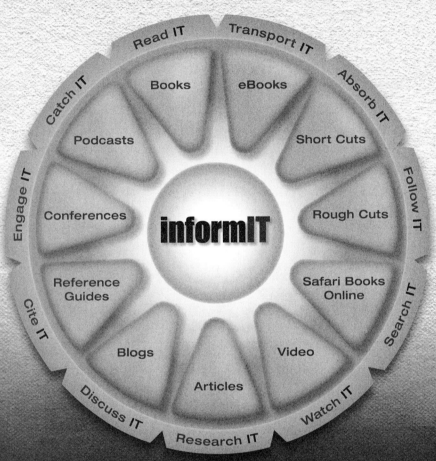

Learn IT at InformIT

Go Beyond the Book

Read IT · **Transport IT** · **Catch IT** · **Absorb IT** · **Engage IT** · **Follow IT** · **Cite IT** · **Search IT** · **Discuss IT** · **Watch IT** · **Research IT**

Books · eBooks · Podcasts · Short Cuts · Conferences · **informIT** · Rough Cuts · Reference Guides · Safari Books Online · Blogs · Video · Articles

11 WAYS TO LEARN IT at **www.informIT.com/learn**

The online portal of the information technology
publishing imprints of Pearson Education

Safari Library
Subscribe Now!
http://safari.informit.com/library

Safari's entire technology collection is now available with no restrictions. Imagine the value of being able to search and access thousands of books, videos and articles from leading technology authors whenever you wish.

EXPLORE TOPICS MORE FULLY

Gain a more robust understanding of related issues by using Safari as your research tool. With Safari Library you can leverage the knowledge of the world's technology gurus. For one flat monthly fee, you'll have unrestricted access to a reference collection offered nowhere else in the world -- all at your fingertips.

With a Safari Library subscription you'll get the following premium services:

- **Immediate access to the newest, cutting-edge books** - Approximately 80 new titles are added per month in conjunction with, or in advance of, their print publication.

- **Chapter downloads** - Download five chapters per month so you can work offline when you need to.

- **Rough Cuts** - A service that provides online access to pre-published information on advanced technologies updated as the author writes the book. You can also download Rough Cuts for offline reference.

- **Videos** - Premier design and development videos from training and e-learning expert lynda.com and other publishers you trust.

- **Cut and paste code** - Cut and paste code directly from Safari. Save time. Eliminate errors.

- **Save up to 35% on print books** - Safari Subscribers receive a discount of up to 35% on publishers' print books.

THIS BOOK IS SAFARI ENABLED

INCLUDES FREE 45-DAY ACCESS TO THE ONLINE EDITION

The Safari® Enabled icon on the cover of your favorite technology book means the book is available through Safari Bookshelf. When you buy this book, you get free access to the online edition for 45 days.

Safari Bookshelf is an electronic reference library that lets you easily search thousands of technical books, find code samples, download chapters, and access technical information whenever and wherever you need it.

TO GAIN 45-DAY SAFARI ENABLED ACCESS TO THIS BOOK:

- Go to **informit.com/safarienabled**
- Complete the brief registration form
- Enter the coupon code found in the front of this book on the "Copyright" page

If you have difficulty registering on Safari Bookshelf or accessing the online edition, please e-mail customer-service@safaribooksonline.com.